T0146260

Military Applications of Artificial Intelligence

Ethical Concerns in an Uncertain World

FORREST E. MORGAN, BENJAMIN BOUDREAUX, ANDREW J. LOHN,
MARK ASHBY, CHRISTIAN CURRIDEN, KELLY KLIMA, DEREK GROSSMAN

Prepared for the United States Air Force
Approved for public release; distribution unlimited

For more information on this publication, visit www.rand.org/t/RR3139

Library of Congress Cataloging-in-Publication Data is available for this publication.
ISBN: 978-1-9774-0310-0

Published by the RAND Corporation, Santa Monica, Calif.

© Copyright 2020 RAND Corporation

RAND® is a registered trademark.

Support RAND

Make a tax-deductible charitable contribution at
www.rand.org/giving/contribute

www.rand.org

Preface

The research in this report was conducted over the course of one year, from October 2017 to September 2018. The completed report was originally delivered to the sponsor in October 2018. It was approved for public distribution in March 2020. Since the research was completed and delivered, new organizations have been created and important steps have been taken to address many of the topics the report describes. As a result, this report does not capture the current state of the topic at the time of publication. Although expert and public opinions may have shifted, we believe the report documents a useful view of perspectives.

The field of artificial intelligence (AI) has advanced at an ever-increasing pace over the last two decades. Systems incorporating intelligent technologies have touched many aspects of the lives of citizens in the United States and other developed countries. It should be no wonder then that AI also offers great promise for national defense. A growing number of robotic vehicles and autonomous weapons can operate in areas too hazardous for human combatants. Intelligent defensive systems are increasingly able to detect, analyze, and respond to attacks faster and more effectively than human operators can. And big data analysis and decision support systems offer the promise of digesting volumes of information that no group of human analysts, however large, could consume and helping military decisionmakers choose better courses of action more quickly.

But thoughtful people have expressed serious reservations about the legal and ethical implications of using AI in war or even to enhance security in peacetime. Anxieties about the prospects of "killer robots" run amok and facial recognition systems mistakenly labeling innocent citizens as criminals or terrorists are but a few of the concerns that are fueling national and international debate about these systems.

These issues raise serious questions about the ethical implications of military applications of AI and the extent to which U.S. leaders should regulate their development or restrain their employment. But equally serious questions revolve around whether potential adversaries would be willing to impose comparable guidelines and restraints and, if not, whether the United States' self-restraint might put it at a disadvantage in future conflicts.

With these concerns in mind, the Director of Intelligence, Surveillance, and Reconnaissance Resources, Headquarters, United States Air Force (USAF), commissioned a fiscal year 2017 Project AIR FORCE study to help the Air Force understand the ethical implications of military applications of AI and how those capabilities might change the character of war. This report, which is one of the products of that study, seeks to answer the following questions: (1) What significant military applications of AI are currently available or expected to emerge in the next 10–15 years? (2) What legal, moral, or ethical issues would developing or employing such systems raise? (3) What significant military applications of AI are China and Russia currently pursuing? (4) Does China, Russia, or the United States have exploitable vulnerabilities due to ethical or cultural limits on the development or employment of military applications of AI? (5) How can USAF maximize the benefits potentially available from military applications of AI while mitigating the risks they entail?

The research described in this report was conducted within the Strategy and Doctrine Program of RAND Project AIR FORCE.

RAND Project AIR FORCE

RAND Project AIR FORCE (PAF), a division of the RAND Corporation, is USAF's federally funded research and development (R&D) center for studies and analyses. PAF provides the Air Force with independent analyses of policy alternatives affecting the development, employment, combat readiness, and support of current and future air, space, and cyber forces. Research is conducted in four programs: Strategy and Doctrine; Force Modernization and Employment; Manpower, Personnel, and Training; and Resource Management. The research reported here was prepared under contract FA7014-16-D-1000.

Additional information about PAF is available on our website:
www.rand.org/paf

This report documents work originally shared with USAF on September 27, 2018. The draft report, issued on October 10, 2018, was reviewed by formal peer reviewers and USAF subject matter experts (SMEs).

Table of Contents

Figures

Tables

Summary

The research in this report was conducted over the course of one year, from October 2017 to September 2018. The completed report was originally delivered to the sponsor in October 2018. It was approved for public distribution in March 2020. Since the research was completed and delivered, new organizations have been created and important steps have been taken to address many of the topics the report describes. As a result, this report does not capture the current state of the topic at the time of publication. Although expert and public opinions may have shifted, we believe the report documents a useful view of perspectives. Furthermore, several chapters present historical perspectives of American, Chinese, and Russian willingness or reticence to develop and/or field artificial intelligence (AI) for the battlefield.

This report examines military applications of AI and considers the ethical implications of employing them in war and peace. AI encompasses a wide range of technologies, many of which offer great promise in military applications. Consequently, the United States, China, and Russia are all developing military applications of AI. Given the rapid progress seen in this field during the last two decades, this could change the character of war in the coming years.

But thoughtful people have expressed serious reservations about the legal and ethical implications of military forces using AI. The most strident objections have revolved around prospects of machines killing people, without direct approval of human operators. But concerns have been raised about other applications of AI as well, such as decision support systems that might select questionable targets without commanders being able to examine the complex calculations behind such choices, and citizens being detained or even killed as a result of a facial recognition system or some other complex AI calculation misidentifying them as terrorists or criminals.

These concerns raise serious questions about the ethical implications of military applications of AI and the extent to which U.S. leaders should regulate their development or restrain their employment. But equally serious questions revolve around whether potential U.S. adversaries would be willing to impose comparable guidelines and restraints and, if not, whether the United States' self-restraint might put it at a disadvantage in future conflicts. With these concerns in mind, this report seeks to answer the following questions for the U.S. Air Force (USAF):

1. What significant military applications of AI are currently available or expected to emerge in next 10–15 years? What benefits do they offer? What operational or strategic risks do they entail?
2. What legal, moral, or ethical issues would developing or employing such systems raise? How sensitive to these issues are U.S. citizens?
3. What significant military applications of AI are China and Russia currently pursuing? Are political or military leaders in those countries interested in pursuing norms or

international agreements that would constrain the employment of autonomous weapons or other military applications of AI?

4. Does China, Russia, or the United States have vulnerabilities due to ethical or cultural limits on the development or employment of military applications of AI? If so, could opponents exploit these vulnerabilities?

5. What do answers to the foregoing questions suggest that USAF leaders need to do to maximize the benefits potentially available from military applications of AI, while mitigating the risks they entail?

Methodology

To answer these questions, the RAND team interviewed 29 experts in the field of AI and other areas relevant to this study. Drawing from insights we gained in these discussions, we developed a 26-question survey (plus nine demographic questions), which we administered online to approximately 2,500 people in the United States, polling their attitudes regarding the ethical acceptability of various military applications of AI across a range of strategic contexts.

While accomplishing these tasks, we reviewed the current and historical literature to better understand the development of AI and its emerging capabilities. We assessed the legal and ethical risk of these capabilities in terms of the law of armed conflict (LOAC) and just-war doctrine. We also surveyed Chinese and Russian AI literature and other sources in order to assess military AI programs and intentions in those countries.[1]

Findings

There Is No Consensus on an Artificial Intelligence Development Timeline, but Experts Agree That There Will Likely Be a Steady Increase in the Integration of Artificial Intelligence in Military Systems

Developments in military AI could cause a dramatic evolution, perhaps even a transformation, in the character of war. Yet the experts we interviewed offered a wide range of opinion regarding when, or even if, such a change might occur. AI technologies have developed rapidly and are being integrated into an increasing number of military applications. The United States, China, and Russia are all vigorously pursuing military AI capabilities. These technological developments have serious ramifications for a wide range of warfighting capabilities. As a result, careful consideration of how these capabilities can be employed in keeping with the LOAC and how the rules of engagement will need to accommodate them in future scenarios is needed. While we cannot predict how quickly military AI systems will emerge, we believe that, based on the pace of developments to date, there will be time to develop these understandings and establish appropriate safeguards if U.S. leaders are proactive.

---

[1] Although this is an unclassified study, we also reviewed classified sources to guard against making erroneous statements about foreign developments. No classified information was included in this report.

The United States Faces Significant International Competition in the Field of Military Artificial Intelligence

Unlike in past technological developments, such as atomic weapons and stealth aircraft, the United States will not have a monopoly, or even a first-mover advantage, in the competition for military AI. China is aggressively developing robotic systems and an assortment of other systems to integrate data from a wide variety of sensors to identify hidden targets, provide a common operating picture to commanders, and enable rapid decisionmaking.

Russia also has an advanced robotics program but is more actively pursuing other areas, such as defensive systems, decisionmaking and planning tools, electronic warfare, cyber warfare, and AI-driven disinformation campaigns. Russia is already using AI technologies in support of its hybrid, gray-zone, and information warfare operations abroad.

The Development of Military Artificial Intelligence Presents a Range of Risks That Need to Be Addressed

Researchers, technologists, and citizens in the United States and elsewhere have all raised concerns about risks associated with military AI. As Figure S.1 indicates, we have surveyed these risks and sorted them into three categories: ethical, operational, and strategic.

Figure S.1. Taxonomy of Artificial Intelligence Risk

Ethical	Operational	Strategic
Law of Armed Conflict Accountability and Moral Responsibility Human Dignity Human Rights and Privacy	Trust and Reliability Hacking, Data Poisoning, and Adversarial Attacks Accidents and Emergent Risks	Thresholds Escalation Management Proliferation Strategic Stability

Each of these risk categories presents serious challenges. Ethical risks are important from a humanitarian standpoint. States are obligated to abide by the provisions of International Humanitarian Law (IHL), which seek to protect innocent civilians from the violence and abuses of war. Autonomous weapons, or those capable of identifying and destroying targets without human operators in the decision cycle, raise fundamental questions about moral responsibility, the protection of human dignity, and whom to hold accountable for harmful action if the wrong targets are attacked. Systems enabling governments to collect and analyze large volumes of personal data raise human rights and privacy concerns. Operational risks, such as those related to the reliability, fragility, and security of AI systems, raise fundamental questions about whether military AI systems will function according to the intent of military commanders and operators. Strategic risks, including the risks that AI will increase the likelihood of war, escalate ongoing conflicts, and proliferate to malicious actors, are important to U.S. leaders and the international community.

China and Russia are not immune to these risks, although they may be less sensitive to some than others. For instance, China has proposed a ban on lethal autonomous weapon systems (LAWS)—which Russia and the United States do not support—but Beijing's proposed ban defines LAWS so narrowly that it would probably not constrain China's development or use of these weapons even if the international community accepted it. This leads us to question whether Beijing's professed concerns about human dignity and moral responsibility are genuine. And China and Russia are clearly less sensitive to some other ethical concerns, such as their citizens' rights to privacy. On the other hand, there is reason to believe that Beijing and Moscow do genuinely care about the operational and strategic risks entailed in military AI. No military or political leader wants lethal weapons that are unreliable, can be hacked, or might exhibit unpredictable emergent behaviors. Nor does any national leader want his or her military commanders advised by decision support systems that might recommend actions that are insensitive to escalation thresholds and thereby risk stability in a crisis or escalation in war. In fact, these concerns might be even greater in China and Russia than in some other countries, given their political and strategic cultures, which emphasize centralized control.

International Competition Could Encourage Countries to Rush the Development of Military Artificial Intelligence Without Sufficient Attention to Safety, Reliability, and Humanitarian Consequences

International competition in the development of military AI could escalate into a full-blown arms race. The lack of international consensus on norms of responsible development and use creates risks that states will have an incentive to rapidly acquire and integrate military AI without putting appropriate policies in place to ensure that systems are safe and reliable. This situation could result in a "race to the bottom," ultimately threatening the ability of humans to exercise agency over military AI systems.

The U.S. Public Appears to Support the Department of Defense's Continued Investment in Military Artificial Intelligence, but the Public's Acceptance of Risk Varies by Context

The results of our survey of public opinion suggest that the U.S. public supports Department of Defense's (DoD's) investment in military AI applications. However, the results also indicate that the public is concerned about the ethical risks that military AI poses for accountability and human dignity. The public appears to hold strong convictions about the importance of human control over the use of autonomous weapons and to believe that an operator should be required to authorize attacks that take human life. Interestingly, despite the respondents' objections to the use of autonomous weapons to kill people, they were more likely to permit it if U.S. forces were losing a battle, especially if the enemy was using autonomous weapons.

Despite Ongoing United Nations Discussions, an International Ban or Other Regulation on Artificial Intelligence in Military Applications Is Not Likely in the Near Term

A significant number of countries supports a new, legally binding treaty that would ban the development and use of autonomous weapons.[2] However, most of the major military powers— including the United States, the United Kingdom, and Russia—see significant value in military AI and do not wish to create new international constraints that could slow its technological development. Given the resistance of these major powers and other states, the international community is not likely to agree to a treaty banning or regulating the development of autonomous weapons or other applications of military AI anytime soon.

There Is Growing Recognition That Risks Associated With Military Artificial Intelligence Will Require Human Operators to Maintain Positive Control in Its Employment

The risks associated with military AI are most serious in cases where systems act autonomously without human direction or oversight over their critical functions. To grapple with these risks, international discussions at the United Nations Conference on Certain Conventional Weapons (UN CCW) have been moving toward consensus that LOAC requires some level of human involvement in all military action.

Despite this developing consensus, it is unclear how states will interpret this requirement in practice or take steps to ensure that military AI systems do not outpace legal and humanitarian restraints. Yet even China and Russia have noted the importance of human operators in exercising some degree of supervision or oversight over military AI systems. And as mentioned above, these states, like the United States and its allies, have national interests in mitigating operational and strategic risks by ensuring human control over military AI and will likely want to ensure that military commanders have control over weapon systems.

Recommendations

This research leads to three recommendations for the Air Force and the Office of the Secretary of Defense (OSD), and three additional recommendations for the Air Force, Joint Staff, and other

---

[2] United Nations Office for Disarmament Affairs, "Pathways to Banning Fully Autonomous Weapons," October 23, 2017; Phil Dierking, "Support Grows for a Treaty to Ban Killer Robots," *VOA Learning English*, August 30, 2018; Mattha Busby and Anthony Cuthbertson, "Killer Robots Ban Blocked by U.S. and Russia at UN Meeting," *The Independent*, September 3, 2018.

DoD entities working in cooperation with the Department of State. Following are recommendations for Air Force and OSD leaders.

- **Organize, train, and equip forces to prevail in a world in which military systems empowered by AI are prominent in all domains.** Although it is impossible to predict how soon military AI will be so capable that it changes the character of war, this research suggests that significant advances will occur in the next 10–15 years. China, Russia, and other state and nonstate actors are aggressively pursuing AI capabilities. The United States must stay at the forefront of military AI capability development. While U.S. leaders must always be cognizant of the dangers and potential costs of an arms race, not to compete in an arena where potential adversaries are developing dangerous capabilities is to cede the field. That would be unacceptable. Instead, military AI development should be pursued with all necessary precautions to mitigate risks and to ensure that appropriate human judgment is applied in all phases of development, testing, and employment. Before LAWS are employed, commanders will need to develop rules of engagement that ensure human control is exercised at levels appropriate to the operational and strategic context of each situation. Professional military education will need to include instruction on the risks and responsibilities of operating AI-empowered military systems. Operators will need to be trained in realistic environments in order to develop the appropriate levels of trust, neither overtrusting nor undertrusting the systems under their control, to avoid automation surprise.

- **Understand how to address the ethical concerns expressed by technologists, the private sector, and the American public.** These stakeholders have genuine and sincere worries about the implications of military AI and the risks that humans will have less agency over life-and-death decisions in war. Recent developments, such as Google's decision to withdraw from Project Maven, suggest that there is a deficit of trust between key stakeholders and the U.S. government regarding military AI. It is important to regain and maintain that trust. To do so, the OSD and Air Force will need to convince these stakeholders that they take their concerns seriously.

- **Conduct public outreach to inform stakeholders of the U.S. military's commitment to mitigating ethical risks associated with AI to avoid a public backlash against "killer robots" and the resulting policy limitations for Title 10 action.** Identify opportunities to speak publicly about the U.S. military's commitment to mitigating the risks of autonomous weapons and other applications of AI. Many elements of DoD policy on military AI are broadly consistent with the demands of arms control advocacy groups and other actors and go a long way toward mitigating the risks they are most concerned about. These policy elements should be publicly underscored. Citizens should be informed that OSD and Air Force development efforts are concentrated in areas where the public is most supportive of military AI, such as force protection, improved compliance with LOAC, and systems intended to improve logistics and manpower issues. More emphasis on this will help OSD and the Air Force build trust in their stewardship of AI systems.

Recommendations for Air Force, Joint Staff, and other DoD leaders working in cooperation with the Department of State are that they

- **Follow discussions at the United Nations Convention on Certain Conventional Weapons and track the evolving positions held by stakeholders in the international**

community. States and advocacy groups have offered formal position statements at the UN CCW regarding autonomous weapons. Russia and China have submitted papers stating their positions. While sometimes cryptic and vague, these statements offer opportunities to discern official state positions on system developments that Beijing and Moscow may otherwise be reticent to discuss. The State Department, OSD, Joint Staff, and Air Force should follow the UN CCW process to better understand these views as well as those of allies and other important stakeholders.

- **Seek greater technical cooperation and policy alignment with allies and partners regarding the development and employment of military AI.** A major advantage the United States enjoys in the international environment is its positive relationships with allies and partners around the world. The United States should engage these states in selected development efforts and coordinate policies regarding military AI. By cooperating with partners, the United States can leverage technical comparative advantages, promote shared understandings, encourage the development of compatible policies, and prepare to operate military AI systems in multinational forces.

- **Explore confidence-building and risk-reduction measures with China, Russia, and other states attempting to develop military AI.** Although it is not clear how sincere Beijing and Moscow are in their humanitarian concerns, they at least claim to care about their commitment to LOAC and to ensuring human control over the critical functions of military AI. These states and others should also be interested in mitigating the operational and strategic risks discussed in this report. OSD, the Joint Staff, and the Air Force, in coordination with the State Department, should work to identify areas where states have common interests regarding military AI and then approach their counterparts at the UN CCW, other international forums, or in bilateral settings and pursue engagement in collaborative activities to mitigate risks and begin the development of confidence-building measures.

Closing Observations

Among the principal concerns motivating this study were questions about whether the United States might be constrained in its development or employment of military AI in ways that China and Russia are not, and what the Air Force needs to do to maximize the benefits potentially available from these systems, while mitigating the risks they entail. The findings in this report address these questions and illuminate a way ahead.

China and Russia are vigorously developing military AI and do not appear to have the same ethical concerns as the United States. Therefore, the Air Force should continue its development of military AI in all areas that support more effective mission accomplishment when cost-benefit analyses indicate that development, production, and measure/countermeasure outcomes will be favorable. Most of these developments will probably focus on nonlethal applications, such as intelligence, surveillance, and reconnaissance (ISR) processing, but advanced weapon systems will be needed as well. Any LAWS developed should be designed to operate semiautonomously—that is, with a human operator "in the loop," manually authorizing each use of lethal force. Some systems will also need to be capable of operating with supervised autonomy (operator "on the loop," able to intervene if necessary), or with full autonomy, as the tactical situation requires. In all cases, these systems must be equipped with failsafe override controls that enable operators to keep their actions within the bounds of commanders' intent and provide commanders oversight and the ability to promptly intervene when necessary.

Employment of these weapons should be done within the constraints of LOAC and the guidelines of just-war doctrine. Rules of engagement should require modes of human supervision that enable adequate levels of discrimination and precaution, given the tactical situation, to ensure that risks to noncombatants are proportionate to the importance of military objectives. In most cases, this will require LAWS to run semiautonomously; however, in some scenarios this will not be practical, and if an adversary begins employing LAWS with full autonomy, U.S. forces should be able to match this escalation, in keeping with LOAC and relevant ethical principles, to assure adequate force protection and mission success. Although surveys indicate that the U.S. public is averse to autonomous weapons taking human life, they also suggest that the public supports further development of military AI and understands the need to match enemy escalation to avoid defeat.

Finally, one of this study's research questions asked whether China, Russia, or the United States has vulnerabilities due to ethical or cultural limits and, if so, could these vulnerabilities be exploited. The results of this analysis did not uncover specific exploitable weaknesses among these states at this time. However, AI technologies are developing at a rapid pace. Given the potential consequences of falling behind, it is vitally important that the United States stay at the forefront of military AI development.

Acknowledgments

We would like to thank the sponsor of our study, Mr. Kenneth Bray, Director of Intelligence, Surveillance, and Reconnaissance Resources, for commissioning this work. His clear direction, enthusiastic engagement, and unwavering support not only allowed us to do our best work but also made the entire process a rewarding experience. We also want to express our appreciation to our action officer, Captain Michael Kanaan, for his responsiveness and expertise. He, too, contributed to this study's success.

We are especially appreciative of the 29 subject matter experts (SMEs) who were so generous with their time in allowing us to conduct lengthy interviews with them. Specifically, our thanks go to Kiril Avramov, Samuel Bendett, Terrance Boult, Miles Brundage, Michael Chase, Tai Ming Cheung, Rebecca Crootof, David Danks, Thomas Dietterich, Jeffrey Ding, James Dubik, Michael Dvorak, Jeffrey Engstrom, Benjamin Fernandes, Edward Geist, Ryan Hefron, Elsa Kania, Robert Latiff, Dara Massicot, David McQueeney, Michael Page, Kevin Pollpeter, Heather Roff, Paul Scharre, Kerstin Vignard, and Rand Waltzman. We also thank several other individuals who have asked not to be named.

We also express our appreciation to Rory Lewis, our consulting expert on artificial intelligence at the University of Colorado, Colorado Springs, and our three quality assurance reviewers: Philip Anton, Antonio DeSimone, and Bonnie Triezenberg. All of them provided us helpful feedback, which made this a better product.

We thank our program director, Paula Thornhill, for her patient guidance and our associate program director, Anthony Rosello, for his help in managing budgetary matters. Finally, we thank Laura Pavlock-Albright for preparing this manuscript.

Abbreviations

A2/AD	anti-access/area-denial
ACS	automated control system
AFRL	Air Force Research Laboratory
AGI	artificial general intelligence
AI	artificial intelligence
APS	active protection system
AUV	autonomous underwater vehicle
C2	command-and-control
CASTC	China Aerospace Science and Technology Corporation
CCP	Chinese Communist Party
CEMA	Cyber-Electromagnetic Activity
CESI	China Electronic Standards Institute
CI	confidence interval
CIWS	Close-In Weapons System
DARPA	Defense Advanced Research Projects Agency
DART	Dynamic Analysis and Replanning Tool
DoD	Department of Defense
EW	electronic warfare
GCAS	Ground Collision Avoidance System
GGE	Group of Governmental Experts
HARM	High-Speed Anti-Radiation Missile
HRW	Human Rights Watch
IHL	International Humanitarian Law
ISR	intelligence, surveillance, and reconnaissance

JADE	Joint Assistant for Deployment and Execution
KGB	Komitet Gosudarstvennoy Bezopasnosti (Committee for State Security)
LAWS	lethal autonomous weapon system
LOAC	law of armed conflict
LRASM	Long Range Anti-Ship Missile
ML	machine-learning
MoD	Ministry of Defense
MTurk	Amazon's Mechanical Turk
NTsUO	Natsional'nyy Tsentr Upravleniya Oboronoy (National Defense Management Center)
OODA	observe, orient, decide, and act
OSD	Office of the Secretary of Defense
PLA	People's Liberation Army
R&D	research and development
RYAN	Raketno-Yadernoye Napadenie (Nuclear-Missile Attack)
SAM	surface-to-air missile
SAPE	Survivable Adaptive Planning Experiment
SME	subject matter expert
SWORDS	Special Weapons Observation Reconnaissance Detection System
TASM	Tomahawk Anti-Ship Missile
UAS	uninhabited aerial system
UAV	unmanned aerial vehicle
UCAV	unmanned combat aerial vehicle
UGV	unmanned ground vehicle
UN CCW	United Nations Convention on Certain Conventional Weapons
UNOG	United Nations Office at Geneva

USAF	United States Air Force
USV	unmanned surface vehicle
VRYAN	Vnezapnoe Raketno-Yadernoye Napadenie (Surprise Nuclear-Missile Attack)
WTO	World Trade Organization
XAI	Explainable AI

1. Introduction

The research in this report was conducted over the course of one year, from October 2017 to September 2018. The completed report was originally delivered to the sponsor in October 2018. It was approved for public distribution in March 2020. Since the research was completed and delivered, new organizations have been created and important steps have been taken to address many of the topics the report describes. As a result, this report does not capture the current state of the topic at the time of publication. Although expert and public opinions may have shifted, we believe the report documents a useful view of perspectives. Furthermore, several chapters present historical perspectives of American, Chinese, and Russian willingness or reticence to develop and/or field artificial intelligence (AI) for the battlefield.

This report examines military applications of AI and considers the ethical implications of employing them in war and peace. AI can be broadly defined as "the capability of computer systems to perform tasks that normally require human intelligence," but there are multiple technologies and computational approaches encompassed within this broad definition, and it is often more useful to discuss specific applications in the context of those technologies.[1]

However broadly defined, the field of AI has advanced at an ever-increasing pace over the last two decades. As a result, technologies using AI have already touched many aspects of the lives of citizens in the United States and other developed countries. Smartphones, mobile mapping and navigation systems, natural language interaction with computers, targeted online marketing, and tailored information campaigns in social media are only a few of the many ways that AI is becoming ubiquitous in daily life. This trend will only increase as self-driving vehicles and other autonomous robotic systems become more accepted and integrated into society.

It should be no wonder, then, that AI also offers great promise for national defense. A growing number of robotic vehicles and autonomous weapons are able to operate in combat zones too hazardous for human combatants. Intelligent defensive systems are increasingly able to detect, analyze, and respond to attacks faster and more effectively than human operators can. And big data analysis and decision support systems offer the promise of digesting volumes of information that no group of human analysts, however large, could consume and thereby help military decisionmakers choose better courses of action more quickly. As a result, the United States, China, Russia, and other advanced military powers are all developing military applications of AI. This could change the very character of warfare in the coming years.[2]

---

[1] Definition from Defense Science Board, *Summer Study on Autonomy*, Washington, D.C., June 2016, p. 5.

[2] For more detailed summaries of these emerging capabilities, see Robert H. Latiff, *Future War: Preparing for the New Global Battlefield*, New York: Alfred A. Knopf, 2017; Paul Scharre, *Army of None: Autonomous Weapons and the Future of War*, New York: Norton, 2018.

But thoughtful people have expressed serious reservations about the legal and ethical implications of military forces using AI in war or even to enhance security in peacetime. The most strident objections have revolved around prospects of machines killing people without direct approval of human operators and, potentially, even without their oversight or ability to intervene if weapons select the wrong targets.[3] But concerns have been raised about other applications of AI as well, such as decision support systems that might urge escalatory actions, or even preemptive attacks, without commanders being able to examine the complex calculations behind such recommendations, or citizens being detained or even killed as a result of a facial recognition system or some other complex AI calculation misidentifying them as terrorists or criminals.[4]

These concerns raise serious questions about the ethical implications of military applications of AI and the extent to which U.S. leaders should regulate their development or restrain their employment. But equally serious questions revolve around whether potential U.S. adversaries would be willing to impose comparable guidelines and restraints and, if not, whether the United States' self-restraint might put it at a disadvantage in future conflicts. This report seeks to answer these questions for the U.S. Air Force (USAF).

Background

Many people attribute the birth of AI to Alan Turing's 1950 essay, "Computing Machinery and Intelligence." In it, Turing, a noted mathematician and pioneer of computer science, posed the question of whether machines would ever be able to think. Then, quickly discarding the question as too vague on definitional grounds—after all, what is thinking, exactly?—he proposed what he called the "imitation game." In what is now commonly referred to as the "Turing test," he described putting a computer in one room and a man in another. The man would pose a series of questions designed to determine whether he was talking to a machine or to another man. Turing asserted that, in time, a computer could be programed to answer questions so well that it would be indistinguishable from a human respondent. But does that mean the computer would actually be able to think? Turing implied that it does not really matter.[5]

While Turing's article is thought-provoking, the term *artificial intelligence* was actually first used as a title for a conference held at Dartmouth College in 1955. The organizers of that event proposed that "every aspect of learning or any other feature of intelligence can in principal be so precisely described that a machine can be made to simulate it."[6] This was a remarkably ambitious

---

[3] For instance, see Mark Guburd, "Why Should We Ban Autonomous Weapons? To Survive," *SPECTRUM*, June 1, 2016.

[4] Max Tegmark, "Benefits and Risks of Artificial Intelligence," Future of Life Institute, n.d.

[5] See A. M. Turing, "Computing Machinery and Intelligence," *Mind*, Vol. 49, 1950, pp. 433–460.

[6] J. McCarthy, M. L. Minsky, N. Rochester, and C. E. Shannon, "A Proposal for the Dartmouth Summer Research Project on Artificial Intelligence," August 31, 1955, published in *AI Magazine*, Vol. 27, No. 4, 2006. ·

claim for that era, but it is not clear whether the conference produced anything of value—a final report was never delivered.[7] And in the years since, progress in AI research has gone through visible cycles of boom, when expected advances kindled surges in funding, and bust, when the failure of those expectations to manifest put the field in disfavor with government and industrial patrons.[8]

Meanwhile, a wide range of military systems became increasingly automated without being associated, at least in most people's minds, with AI. For instance, as early as the 1940s, some aircraft and air defense radars were equipped with transponders by which radar operators—and eventually the radar systems themselves—could interrogate the aircraft they were tracking to determine whether they were friendly or hostile.[9] In later years, tactical and strategic warning systems were designed to identify aircraft and missiles by matching the speeds and shapes of their radar returns, or the intensities of their heat signatures, to databases of known threats. By the 1970s, surface-to-air and air-to-air missiles were able to automatically correct course and home in on their targets by radar guidance or with heat-seeking sensors. Over the next couple of decades, air defense systems became increasingly sophisticated and were able to recommend engagement decisions or even engage targets without human intervention if placed in the automatic fire mode. Nevertheless, such systems, though increasingly autonomous, did not give rise to anxiety about "killer robots" run amok, because the scope of their decisionmaking capabilities was so narrow.[10]

However, significant breakthroughs in AI research and development (R&D) began occurring in the late 1990s, and the pace of advances in this field has been accelerating in the years since. Perhaps the first milestone that captured widespread public attention was in 1997, when IBM's intelligent system, Big Blue, defeated then–world chess champion Gary Kasparov in a six-game match. Even more impressively, in 2016, Google DeepMind's AlphaGo system defeated Lee Sedol, the world's top-ranked player of the Asian game of Go, four games to one.[11] This was far more than an incremental advance in capability. While chess has 20 possible first moves per side and 10^{120} total possible board configurations, Go has 361 possible first moves per side and 10^{170} total possible board configurations—reputedly more than the total number of atoms in the universe.[12] This advance reflects a fundamental difference in approach to the development of

---

[7] Jerry Kaplan, *Artificial Intelligence: What Everyone Needs to Know*, Oxford, UK: Oxford University Press, 2016, p. 15.

[8] Kaplan, 2016, p. 16.

[9] Lord Bowden, "The Story of IFF (Identification Friend or Foe)," *IEE Proceedings*, Vol. 132, Pt. A, No. 6, October 1985, pp. 435–437.

[10] To better appreciate the degree of autonomous engagement capabilities present in air defense systems by the mid-1990s, see "The Cooperative Engagement Capability," *Johns Hopkins APL Technical Digest*, Vol. 16, No. 4, 1995, pp. 377–396.

[11] "Google Achieves AI 'Breakthrough' by Beating Go Champion," *BBC News*, January 27, 2016.

[12] Danielle Muolo, "Why Go Is So Much Harder for AI to Beat Than Chess," *Business Insider*, March 10, 2016.

AI in recent years. Whereas early AI focused on programming and computational complexity, current approaches focus on machine-learning. Regarding the former, the Dartmouth Conference organizer, John McCarthy, and coauthor, Patrick J. Hayes, stated that "we regard the construction of intelligent machines as fact manipulators [in contrast to today's machine learning systems] as being the best bet both for constructing artificial intelligence and understanding natural intelligence."[13] Recent developments suggest that McCarthy and Hayes were mistaken.

In the years since Big Blue's triumph, AI research has made dramatic advances in the fields of computer vision, speech recognition, natural language processing, and robotics. Efforts to develop computer vision began in the 1960s but made little headway until about a decade ago, when the application of convolutional neural networks enabled vision-processing systems to "learn" by building models of objects based on observations of large collections of examples.[14] Recent progress has resulted in ever-more reliable facial recognition systems and emerging capabilities for analyzing video data. Speech recognition has, in some ways, been an even more difficult problem, given the many complexities of language. However, the development of hidden Markov modeling—a statistical technique that calculates probabilities regarding the meanings of patterns of sound—and the more recent application of a deep-learning method called "long short-term memory" have enabled advances leading to the speech recognition systems we now have in smartphones and other computer devices.[15] These advances and others allow people to interact with intelligent systems—that is, to enter data, ask questions, and receive spoken or written responses using natural language, rather than computer code.

The most publicized advances in autonomous robotics have probably been in the field of self-driving vehicles. Motivated by prospects for military applications of these capabilities, the U.S. Department of Defense (DoD) has been instrumental in promoting their development. In 2004, the Defense Advanced Research Projects Agency (DARPA) held its first contest in this area, the "DARPA Grand Challenge," in which it offered a $1 million prize to the first self-driving vehicle to cross the finish line in a 142-mile race through rugged terrain in the Mojave Desert. Unfortunately, none of the contestants made it farther than seven and a half miles.[16] Undeterred, DARPA held a second contest the following year, and this time five of the 20

---

[13] John McCarthy and Patrick J. Hayes, "Some Philosophical Problems from the Standpoint of Artificial Intelligence," Stanford University, 1969, p. 4.

[14] A convolutional neural network is a class of deep neural network specialized for analyzing visual imagery. A deep neural network is a complex system of layered algorithms, with the outputs of each layer constituting inputs for subsequent layers; it is inspired by the function of biological neural networks that comprise animal brains. In convolutional neural networks, the connectivity pattern between artificial neurons resembles the organization of the animal visual cortex. See Kaplan, 2016, p. 54.

[15] Sepp Hochreiter and Jurgen Schmidhuber, "Long Short-Term Memory," *Neural Computation*, Vol. 9, No. 8, November 15, 1997, pp. 1735–1780.

[16] Defense Advanced Research Projects Agency (DARPA), "The DARPA Grand Challenge: Ten Years Later," Washington, D.C., March 13, 2014.

entrants completed the course, with a team from Stanford University taking a $2 million prize.[17] Encouraged by this outcome, DARPA held a third contest in 2007, the "Urban Challenge," a 90-kilometer race on city streets, with all the trials that such an environment entailed: stop signs, traffic lights, and requirements to merge with and pass other traffic. This time a team from Carnegie Mellon University won.[18]

These developments, when integrated with previously mentioned advances, such as computer reasoning, image recognition, and precision-guided munitions, raise possibilities for rapid progress in the development of military applications of AI. And indeed, such innovations are now emerging. Big data processing of intelligence, surveillance, and reconnaissance (ISR), sophisticated decision support systems, robotic combat vehicles in all domains, and autonomous weapons are all now within reach; some applications are already available. One of the latest capabilities to emerge is robotic swarming, in which large numbers of autonomous vehicles or weapons are programmed with rules that, when applied in aggregate by the entire group, can exhibit effects of scale and some emergent behaviors that make them much more effective in combat than would be possible by the same number of devices under human control.[19]

These capabilities are welcome developments in the eyes of many military operators. They offer prospects for dramatic increases in combat power and thus the ability to accomplish mission objectives faster and with less exposure to lethal threats. However, they also raise the serious questions that this report has set out to answer.

Purpose and Scope

With the United States, China, and Russia perched on the verge of what might be revolutionary advances in military applications of AI, USAF leaders need to better understand the ethical and legal implications of employing these weapons. But they also need to appreciate the risks of not employing them if potential adversaries choose to do so. To shed light on these important issues, this report seeks to answer the following questions:

1. What significant military applications of AI are currently available or expected to emerge in the next 10–15 years?[20] What benefits do they offer? What operational or strategic risks do they entail?

---

[17] Thomas G. Goodwin and Don Shipley, "Robots Conquer DARPA Grand Challenge," DARPA News Release, October 8, 2005.

[18] John Voelcker, "Autonomous Vehicles Complete DARPA Urban Challenge," *SPECTRUM*, November 1, 2007.

[19] Paul Scharre and Shawn Brimley, "20: The Future of Warfare," *War on the Rocks*, January 29, 2014; Scharre, 2018, pp. 17–22; Kaplan, 2016, pp. 49–53.

[20] Note that this study is concerned with the near-term risks of military applications of AI, as opposed to long-term, potentially existential threats that some people maintain will result from the development of artificial general intelligence (AGI). For information on the potential risks of AGI, see Eliezer Yudkowsky, "Artificial Intelligence as a Positive and Negative Factor in Global Risk," in Nick Bostrom and Milan M. Ćirković, eds., *Global Catastrophic Risks*, New York: Oxford University Press, 2011, pp. 308–345.

2. What legal, moral, or ethical issues would developing or employing such systems raise? How sensitive to these issues are U.S. citizens?
3. What significant military applications of AI are China and Russia currently pursuing? Do political or military leaders in those countries indicate they are concerned about any operational, strategic, ethical, or legal risks of employing such systems? Are they interested in pursuing norms or international agreements that would constrain the employment of autonomous weapons or other military applications of AI?
4. Does China, Russia, or the United States have vulnerabilities due to ethical or cultural limits on the development or employment of military applications of AI? If so, could opponents exploit these vulnerabilities?
5. What do answers to the foregoing questions suggest that USAF leaders need to do to maximize the benefits potentially available from military applications of AI, while mitigating the risks they entail?

Methodology

To answer these questions, we first reviewed the current and historical literature to better understand the development of AI and its emerging capabilities. We also did a preliminary assessment of the legal and ethical risks these capabilities could present in terms of the law of armed conflict (LOAC) and just-war doctrine. With that groundwork laid, we developed structured interviews, which we conducted with 24 experts in the field of AI and other relevant areas. We informally interviewed five other experts as well.[21] Collectively, the interviewees included military developers and operators, thought leaders in business and academia, former officials from DoD, Chinese and Russian area specialists, and retired Air Force and Army general officers. The interviews enabled us to assess what experts believe about the future of military applications of AI in the countries under examination, whether and how soon AI might change the character of war, and what concerns, if any, they have about the legal, ethical, operational, and strategic risks of military AI. While accomplishing these tasks, we surveyed Chinese and Russian AI program developments, reviewing publications from those countries as well as English-language sources assessing military AI programs and intentions there.

Drawing from insights we gained in the foregoing tasks, we developed a 26-question survey (plus nine demographic questions), which we administered online to approximately 2,500 people in the United States, polling their attitudes regarding the ethical acceptability of various military applications of AI in a variety of situations. We did this to assess what policies Air Force leaders and those in other U.S. military and government institutions will need, and how they will need to explain those policies to U.S. citizens, in order to maintain public support for developing and employing military applications of AI.

---

[21] The additional interviews were less formal because they were done on a "target-of-availability" basis, either before the structured interview protocol was finalized, or after the formal interviews were.

Finally, we collated this information, drew findings and observations, and developed recommendations for USAF, other entities within DoD, and the State Department.[22]

Organization

The following chapter explains the basic concepts of AI and describes it military applications. It also addresses approaches to this technology's control and oversight and then discusses the recent technological developments most relevant to its military applications. Chapter 3 explores the risks of developing and employing military AI from a legal and ethical perspective, as well as in terms of operational and strategic risks these systems might also create. Chapters 4, 5, and 6 examine military AI developments in the United States, China, and Russia, respectively. Each chapter provides a brief history of AI development in that country, summarizes what is known about current and future capabilities, and discusses what policies are in place or being proposed to mitigate risks. Chapter 7 discusses U.S. public attitudes regarding the ethics of using military AI. Reporting the results of our public survey, it assesses how U.S. citizens feel about the benefits and risks of employing these capabilities in a range of situations. Finally, Chapter 8 offers findings and recommendations.

In addition, this report has two appendices. Appendix A provides detailed information on the structured interviews we conducted. It includes the interview protocol, the vignettes we used to set context for questions regarding ethical, operational, or strategic risk, and the data derived from an analysis of the interview responses. Appendix B provides information about the public survey we conducted. It lists the survey questions, provides the raw data from the responses, and shows the statistical analysis we used to interpret these data.

Let us turn now to Chapter 2, the military applications of AI.

---

[22] Although this is an unclassified study, we also reviewed classified sources to guard against making erroneous statements about foreign developments out of ignorance. No classified information was included in this report.

2. The Military Applications of Artificial Intelligence

AI has driven significant economic progress in recent years, and as ever greater levels of investment and talent enter the field, more progress should be expected. The number of applications of AI is bound to increase in the future. Businesses and academia have led the development of AI to date. Military applications have lagged behind, but as these technologies mature, they will be employed in an increasing number of military systems. Military establishments around the world are keenly eyeing new developments, hoping that this transformative technology might help them overcome their shortcomings or provide a new form of overmatch.

Perhaps the bluntest appraisal of the implications of AI has been offered by Russian President Vladimir Putin. In a speech to students in September 2017, Putin said, "Artificial intelligence is the future of not only Russia, but of all mankind," and "whoever becomes the leader in this sphere will become the ruler of the world."[1] He was not referring directly to the military applications of AI at that time, but given the pervasive nature of the technology and recent Russian developments, it is wise to conclude that, although the implications of AI will probably be greatest in the economic sphere, the military applications will also be substantial.[2]

This chapter provides an overview of the military applications of AI. It explains the relationship between AI and autonomous systems and discusses approaches to the control and oversight of military applications of this technology. Next, it describes some of the main applications of AI and discusses the benefits they are expected to offer along with the risks they present. Finally, the chapter closes with a consideration of the path ahead for military applications of AI.

What Is Artificial Intelligence?

Artificial intelligence is a term that has been derided for decades. Herbert Simon, one of the founders of the field, expressed his discontent with the grandiose imagery that the term evokes, but conceded, "At any rate, 'artificial intelligence' seems here to stay. . . . In time it will become sufficiently idiomatic that it will no longer be the target of cheap rhetoric."[3] While to some

---

[1] Radina Gigova, "Who Vladimir Putin Thinks Will Rule the World," *CNN*, September 2, 2017.

[2] We will discuss some of these developments in Chapter 7.

[3] Herbert A. Simon, *The Sciences of the Artificial*, 3rd ed., Cambridge, Mass.: MIT Press, 1996, p. 4.

extent that has become true in popular parlance, in technical circles, where more precise definitions are sought, that does not seem to be the case.[4]

The problem with the term is not simply its grandiosity, which sets unreasonable expectations and implies more capability than has historically existed. It also has to do with the shifting nature over time of technologies and capabilities it describes. What would once have been the pinnacle of AI, such as tax-filing software or chess-playing computers, were incremental steps toward more general AI. Such systems are now no longer commonly referred to as "artificial intelligence." That apparent inconsistency is captured in one of the most commonly accepted definitions considered in the Defense Science Board's Summer Study on Autonomy in 2016: "the capability of computer systems to perform tasks that normally require human intelligence"[5]

By that definition, once a technology is common enough that the task it performs no longer requires human intelligence, it ceases to be AI. This seems to agree with common parlance and also explains why the things that were once called AI are now just computing. Further, it uses the very broad phrasing of "perform tasks," which allows AI to encompass the full range of tasks that can be performed. This is what leads to claims such as "AI is the new electricity," but also explains many experts' trepidation in trying to capture AI in a concise definition that provides explanatory or taxonomic value.[6] Instead, they prefer to speak of AI in terms of the applications that it enables, and, in large part, that implies various forms of autonomy.

With these considerations in mind, the Defense Science Board's definition, "the capability of computer systems to perform tasks that normally require human intelligence," is adequate *for the purposes of this document* when discussing AI in very general terms. However, we will usually be referring to AI in the context of specific applications, levels of autonomy, or classes of technology, which we explain in the sections below.

Autonomy and Automation

An important distinction that should be made when discussing applications of AI is whether a system is truly *autonomous* or merely *automated*. When people want a task done, they do it themselves or delegate it to another entity, which can be a human or a machine. In delegating, they give up some control over how it is done, and the entity performing the task has some degree of autonomy. If the task is perfectly scripted with a set of specified and known rules, then technologists say the entity performing it has "low autonomy" and describe it as "automated." If the entity performing the task is empowered to proceed without rules or boundaries, it is

---

[4] It was striking how averse the experts we interviewed were to providing definitions of artificial intelligence.

[5] Defense Science Board, 2016, p. 5.

[6] Shana Lynch, "Andrew Ng: Why AI Is the New Electricity," *Insights by Stanford Business*, Stanford Graduate School of Business, March 11, 2017.

described as fully "autonomous." Nearly all tasks that machines perform fall somewhere between these two extremes, so it makes sense to discuss applications of AI in terms of degrees or levels of autonomy.

Further, people working in the field of AI often distinguish between what they describe as *autonomy-at-rest* and *autonomy-in-motion*.[7] Autonomy-at-rest describes systems that operate in software, or in the virtual world, whereas autonomy-in-motion describes systems that interact largely with the physical world. Examples of autonomy-in-motion that have generated international public concern include lethal autonomous weapon systems (LAWS). Once launched, such weapons can loiter in a designated area of operations for some period of time, hunting for targets. When they identify a target, they attack and destroy it without any human in control of the engagement. This is troubling to many citizens and can give the impression that autonomy-at-rest is safer than autonomy-in-motion, because the risks of autonomy-at-rest are confined to the virtual world. Unfortunately, that impression may be misplaced when military applications of AI are involved. Critical decisions based on autonomy-at-rest can lead to the use of kinetic force with dramatic consequences. In an increasingly digital world, autonomy-at-rest can have increasingly profound effects.

For instance, decisions made based on the algorithmic processing of intelligence could lead to kinetic strikes on the wrong targets in the fog of war. In such cases, a human might execute the strikes, but an AI system could play a substantial role in informing decisionmakers whether to attack and what targets to strike. Senior leaders will likely want human judgment applied in such decisions, and they may think they are getting it. But given the hierarchical nature of military decisionmaking, recommendations from below might be informed by autonomous intelligence processing without the knowledge of individuals actually making the decisions. Autonomy-at-rest can result in "humans-in-motion," which can have lethal, potentially catastrophic, outcomes.

With AI encompassing so many kinds of systems and levels of autonomy, it is helpful to classify these technologies in a graphic taxonomy illustrating relationships between them. Figure 2.1 provides such a taxonomy.

As the figure illustrates, systems generally described as AI encompass a wide range of technologies with varying degrees of complexity and sophistication. As mentioned in Chapter 1, early approaches to AI involved developing automated systems with the ability to perform scripted tasks according to sets of specified rules. Such approaches are still used to some extent, but over the last couple of decades, more sophisticated systems capable of machine-learning (ML) have been developed. These systems can progressively improve their performance by recognizing patterns in large volumes of data and taking corrective actions to improve their

---

[7] Defense Science Board, 2016, p. 5.

Figure 2.1. Taxonomy of Artificial Intelligence Technologies

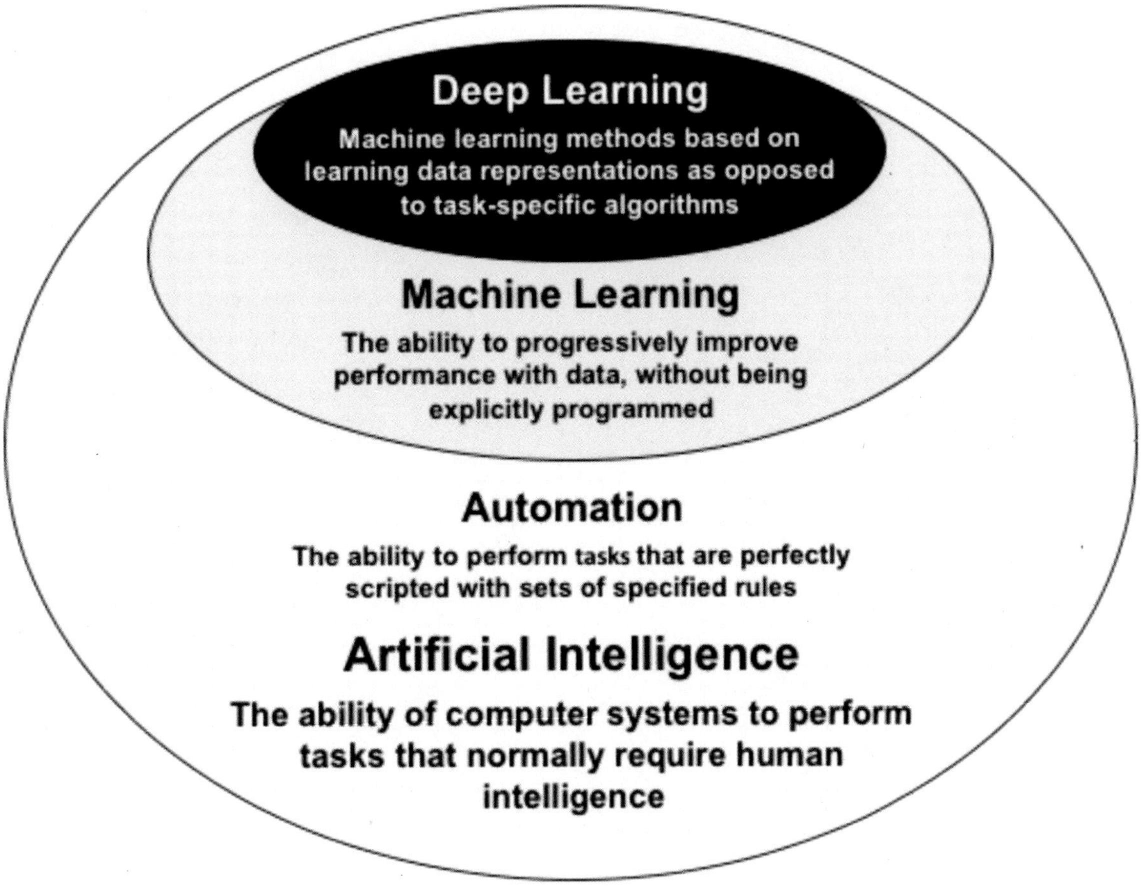

abilities to classify future patterns without being explicitly programmed to do so.[8] An even more sophisticated class of ML systems exhibit deep learning. They use multilayered artificial neural networks to recognize patterns in data representations, such as labeled images, as opposed to using task-specific algorithms as is done in more basic ML systems.[9] As we shall discuss below, recent breakthroughs in deep learning using deep neural networks have enabled significant advances in computer vision and image recognition systems.

Approaches to Control and Oversight

Questions about who or what is ultimately in control of decisions and actions regarding the use of lethal force and who or what is ultimately responsible for the consequences may appear straightforward, but they can be complex and context-dependent. The first issue that arises is how to describe the degree of human involvement in such situations. Developers tend to sort

---

[8] Kaplan, 2016, pp. 27–28.

[9] Kaplan, 2016, p. 34.

human-machine relationships of this nature into three broad categories: "human-in-the-loop," "human-on-the-loop," and "human-out-of-the-loop," or alternatively "semiautonomous," "supervised autonomous," and "fully autonomous."

The loop to which this refers is the *observe, orient, decide,* and *act* (OODA) loop, a concept developed by USAF Colonel John Boyd in the 1980s, which has since become a central tenet in the military doctrine of the United States and other Western countries.[10] In this conception, the *observe* and *orient* components refer to detecting and identifying targets, while the *decide* and *act* components refer to engaging and possibly destroying those targets. The objective is to "get inside the enemy's decision cycle"—that is, to complete the OODA loop and destroy enemy combatants before they can complete their own OODA loop and attack or escape.[11] It is the decision-to-engage part of this process that is so critical, since it concerns whether to use force to kill human beings—hence, the concern with whether humans are "in," "on," or "out of" the loop.

A human is "in-the-loop" when an operator is required to make a positive decision to engage a target. The weapon might observe and orient itself autonomously—finding and identifying enemy targets and queuing them up for engagement—but without an explicit human authorization, the weapon will not engage; it is only semiautonomous. A human is "on-the-loop" when the weapon can autonomously find, identify, and engage targets without human interaction, but an operator is monitoring the situation and has the ability to intervene to prevent or discontinue the engagement. This is supervised autonomy. Conversely, if human operators are "out-of-the-loop" and do not have the ability to intervene in the engagement, the weapon is fully autonomous.

While there are significant legal and moral implications in the differences between these configurations, there may be little or no difference in the engineering of the weapons. In many cases, there is very little technological difference between a weapon system that is supervised autonomous and one that is fully autonomous. It may be only a regulatory or procedural change that is required to convert one to the other. Indeed, in cases where the OODA loop cycles too quickly for a human to meaningfully intervene, a weapon designed to be a supervised autonomous system may, in fact, be fully autonomous in its employment. It may also be a simple technical matter to convert a semiautonomous system to either a supervised autonomous system or a fully autonomous system. It may require only the push of a button or a software update.[12]

This becomes an important consideration, not only in terms of safety and development strategies but also in terms of international relations and conflict. The experts interviewed in this study opined that there may be ethical restraint around the use of autonomous weapons and AI in

---

[10] Frans Osinga, *Science, Strategy, and War: The Strategic Theory of John Boyd*, Delft: Eburon Academic Publishers, 2005, pp. 4–5.

[11] Osinga, 2005, p. 8.

[12] We should point out that while this may be easy from a technical perspective, it can create system-level safety issues. It is often difficult to ensure safety unless proper controls are designed very carefully from inception. See Nancy G. Leveson, *Engineering a Safer World: Systems Thinking Applied to Safety*, Cambridge, Mass.: MIT Press, 2011.

warfare more broadly if the adversary is not using these technologies. Conversely, if they are using these technologies, then the ethical restraint may be loosened. That sentiment also appeared in the public opinion survey we conducted where respondents were substantially more approving of the use of autonomous weapons if the adversary was using autonomous weapons. However, given the ease of switching between semiautonomous, supervised autonomous, and fully autonomous modes, and the lack of transparency in the command-and-control (C2) of these systems, it may be difficult to know what level of autonomy an adversary is using in any particular engagement.[13]

Recent Progress in Artificial Intelligence That Is Driving Military Applications

Technological progress in the development of AI is being driven largely by business demands for many reasons. These include the ability to leverage capital investments and academic resources that are difficult or impossible for military services to provide. As a result, we can reasonably expect most future military applications to be adaptations of technologies developed in the commercial sector. There, progress is coming on a variety of fronts. The next few subsections will briefly outline a few of the most important areas of technological progress in AI that have potential application in military contexts.

Image Recognition

It can be surprising at times to learn which tasks are difficult for computers to perform and which are easy. An observation known as *Moravec's paradox* highlights how tasks that are easy for people are often hard for computers and vice versa.[14] One example of this paradox is image recognition. Humans recognize images easily and unconsciously, but for many years it was a challenge well beyond the capability of computers.[15]

Recently, due to progress primarily in deep neural networks, computer vision has experienced a step-transition in image recognition and object detection and now exceeds human ability in some of these tasks, although it can still exhibit surprising failures compared with humans.[16] Advances in this area have also been driven by the proliferation of images, and particularly

---

[13] As a comparison, consider the fact that proving an actor has used chemical weapons can be very contentious despite the availability of physical evidence. The use of varying levels of AI or autonomy could generate very little evidence.

[14] Hans Moravec, *Mind Children*, Cambridge, Mass.: Harvard University Press, 1988, p. 15.

[15] Bo Zhang, "Computer Vision vs. Human Vision," *IEEE International Conference on Cognitive Informatics*, July 2010.

[16] Kaiming He, Xiangyu Zhang, Shaoqing Ren, and Jian Sun, "Delving Deep into Rectifiers: Surpassing Human-Level Performance on ImageNet Classification," *Proceedings of the 2015 International Conference on Computer Vision*, December 2015.

labeled images, now available. Deep neural networks are "trained" by being exposed to high numbers of labeled images. The progress extends to both facial recognition and more subtle aspects of facial expression, making technology for biometric identification through facial features and even emotional analysis viable.

From a military standpoint, the ability to detect objects and recognize images, and especially to recognize faces and perform emotional analysis, has clear applications. We shall return to this later in this chapter.

Text Analysis

Applying the recent progress in deep neural networks to the vast quantities of digitally written data now available has also resulted in rapid advances in several aspects of natural language processing. Although human-level performance in language-heavy tasks is proving to be more elusive than it was for imagery, machine translation has become a viable option in a growing number of applications and contexts.[17] There have also been substantial advances in applications such as summarization and sentiment analysis, not to mention search engines, although much of the progress in these fields has been made using approaches other than deep neural networks.[18]

Self-Driving Cars

Significant progress has been made in the development of self-driving cars in recent years. In 2004, when DARPA hosted its first Grand Challenge, driverless vehicles were tasked to traverse 142 miles in the desert. The farthest any of them traveled before failing was only 7.5 miles, and no competitor was declared to have won.[19] Today, autonomous vehicles are being tested on the road in several cities around the world. They are still controversial and have been involved in several fatal accidents, but they are beginning to earn the trust of drivers and passengers.

Currently, these systems require humans to be available to intervene much the way pilots are required when aircraft are on autopilot; using the terminology of weapon systems, we would refer to them as "supervised autonomous systems." There is still intense public debate about the safety of these systems. The evolution of this debate could be telling, both about the effectiveness of supervision as a safety mechanism and in how blame might be assigned in accidents involving supervised autonomy.

---

[17] Quoc V. Le and Mike Schuster, "A Neural Network for Machine Translation, at Production Scale," *Google AI Blog*, September 27, 2016.

[18] Mahak Gambhir and Vishal Gupta, "Recent Automatic Text Summarization Techniques: A Survey," *Artificial Intelligence Review*, Vol. 47, No. 1, January 2017, pp. 1–66; Erik Cambria, "Affective Computing and Sentiment Analysis," *IEEE Intelligent Systems*, Vol. 31, No. 2, March 2016, pp. 102–107.

[19] DARPA, 2014.

Game-Playing

Another common benchmark for measuring progress in AI throughout its history has been the ability to play games. When IBM's Deep Blue famously beat Gary Kasparov at chess in 1997, the event was considered a crowning achievement in AI. A similarly momentous and surprising event occurred when Google DeepMind's AlphaGo beat Lee Sedol at Go in 2016.[20] In terms of the complexity of board positions and number of possible moves, Go is a far more difficult game than chess, so Go enthusiasts, and even AI researchers, thought it would be another decade before computers would achieve superhuman performance in that game.[21]

In other regards, though, chess and Go are relatively simple games. The rules are clearly defined, the moves are sequential, and both players have full information about the state of the game. A greater challenge is presented in games such as poker, where players have only partial information and bluffing is involved. These games have been out of reach of computers for many years, but AI systems have recently beaten professional poker players.[22] And now an effort is underway to develop AI that can play games such as StarCraft, which involve several tiers of strategic action over an extended timeline with many more moves than the typical Go game or poker hand.[23]

These game-playing AI systems tend to rely on an approach called *reinforcement learning*, which is not exactly new but has rapidly been gaining popularity. Whereas for image recognition, one typically needs a large set of images and their associated classifications, for reinforcement learning, what is needed is to assign scores or "rewards" to the AI "agent" as it makes decisions, transitioning its environment from one state to another. The goal of the AI agent is to accumulate the greatest rewards possible. This technique can also be applied outside of game-playing to, for instance, robot motion.[24]

But despite these advances, applying game-playing AI to abstract military strategy or wargaming is still very aspirational.

Benefits of Artificial Intelligence in Warfare

Benefits of AI in warfare are often assumed but not explicitly stated. To identify the potential benefits of military applications of AI, we asked the experts we interviewed to name them. All of

[20] Cade Metz, "In Two Moves, AlphaGo and Lee Sedol Redefined the Future," *WIRED*, March 16, 2016.

[21] Cade Metz, "In a Huge Breakthrough, Google's AI Beats a Top Player at the Game of Go," *WIRED*, January 27, 2016.

[22] Tonya Riley, "Artificial Intelligence Goes Deep to Beat Humans at Poker," *Science*, March 3, 2017.

[23] Yoochul Kim and Minhyung Lee, "Humans Are Still Better Than AI at StarCraft—for Now," *MIT Technology Review*, November 1, 2017.

[24] Sergey Levine, Peter Pastor, Alex Krizhevsky, and Deirdre Quillen, "Learning Hand-Eye Coordination for Robotic Grasping with Deep Learning and Large-Scale Data Collection," *International Journal of Robotics Research*, Vol. 37, Nos. 4–5, 2017, pp. 421–436.

the interviewees were able to suggest at least one benefit, and most provided several. Binning similar responses, Figure 2.2 shows the counts for the number of benefits interviewees suggested in each category. Each of these categories will be discussed briefly in this section.

Figure 2.2. Potential Benefits of Military Applications of Artificial Intelligence Identified in Structured Interviews

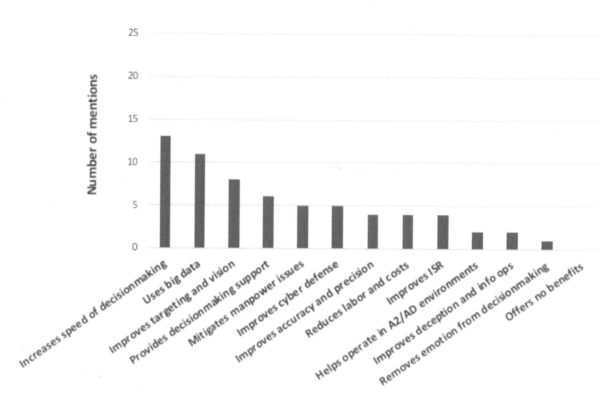

Speed of Decisionmaking

The most frequently mentioned category of benefits of AI in warfare is speed. Interviewees often discussed this in reference to the OODA loop, with the idea being that if it is possible to cycle through the OODA loop faster than one's adversaries, then they will be unable to perform the counteractions needed to defend against one's attacks or to generate their own offensive options fast enough to outpace counteractions.

There are certainly cases where this type of advantage can be envisioned; however, it is also important to keep in mind that timelines are not always dominated by the decision processes that AI can help accelerate. Often the timelines are dominated by the time it takes to move equipment or people or even just the time that munitions are moving to targets. It is important not to overstate the value of accelerating the decision process in these cases.

It is also worth considering whether accelerating decision timelines introduces new risks or aggravates existing ones. There are situations where providing the adversary with more time may make it more likely to select an option that is favorable to the United States. A standoff in times of crisis where it is hoped that the adversary will back down or offer to negotiate could be an example of a high-stakes situation in which more time could be advantageous as opposed to less.

A further risk is that if speed is made the priority attribute for selecting between competing autonomous weapon systems for development, safeguards and robustness might be sacrificed, resulting in weapons that are less safe or reliable than they could be. Despite these caveats though, there is a clear military benefit to the increased speed that AI could provide, as indicated by the large number of experts who suggested it.

Use of Big Data

Big data has become a bit of a catch-all term, and to some degree it was used as such by our expert interviewees. However, there are several concrete concepts that can help to clarify what is meant by this term. It is usually used to describe data that are sized too large to be stored on a computer's memory, are generated too quickly to be managed by a single computer, or take many different forms or formats. Because of those three issues, it can be challenging for humans to make sense of the information that is contained in the data, but machines and AI tend to perform better the more data made available to them.

Improved performance is one of the benefits that the interviewees had in mind when they discussed the use of big data. Additionally, however, they pointed out that the sheer volume of information being collected by various sensors is more than a human or team of humans can analyze. Given the ever-growing volume of data available in the world today, AI is expected to continue to increase in prominence.

Improved Targeting and Vision

One of the areas where data overload is felt most acutely is in image-processing. The number of cameras conducting surveillance in domestic and foreign environments has increased dramatically and is expected to continue to do so. With all the data being generated, there is a clear need and motivation for automation in the process of analyzing incoming video and imagery.

As discussed earlier in this chapter, automated image-recognition and object-detection capabilities have surpassed human ability in at least some cases. As progress continues, these systems will increasingly be able to identify objects that humans would miss. This is already the case for AI that detects skin cancer from images, and it is not unreasonable to expect the same from AI for military or counterterrorism applications.[25] Further, the progress in facial recognition could be applied for quickly identifying terrorists or known combatants, and facial expression analysis could help alert soldiers and other security personnel to risky situations or better manage social interactions while building the peace.

---

[25] Emily Price, "AI Is Better at Diagnosing Skin Cancer Than Your Doctor, Study Finds," *Fortune*, May 30, 2018.

Decisionmaking Support

Driven largely by the progress observed in gaming systems and personal assistant technology, AI is anticipated to be able to recommend options to decisionmakers more quickly, or in some cases, to be able to provide superior options to select from than humans could offer. A familiar example is routing technology that can ingest complete maps and real-time or projected traffic information in ways that humans would not be able to. There are natural applications of such capabilities in logistics, and they are also expected to help with other common tasks, such as scheduling.

More ambitiously, some believe that progress in games such as Go, poker, and StarCraft indicates potential for applying AI in strategic planning tasks. Even if these technologies are not appropriate for use in making combat suggestions or decisions, experts anticipate that they could be used to provide a wider range of possible adversary actions in wargaming and red-teaming events (testing messages, actions, or strategies on exercise participants who have studied a potential adversary's typical behavior in an effort to anticipate how that adversary might react in the real world), or to provide blunder-detection assistance. Blunder avoidance is one of the benefits of using computer teammates in chess, because as circumstances grow complex, it is easy for humans to forget about aspects of the problem or implications of their actions. Computers can be capable of warning about actions that are predictably suboptimal. Although the complexity faced in the real world far exceeds that of the constrained game of chess, some of the experts we interviewed were hopeful that these benefits will be realized.

Mitigation of Manpower Issues

The military has a variety of unmet needs or latent demands for which there are simply not enough personnel. Of course, there is always interest in maintaining the capacity of the force, but there is also an often-discussed gap between demand and personnel available for tasks such as image analysis and foreign language translation. These are the types of tasks that arise from the rapid growth in the volume of data available for processing. Fortunately, they are the types of tasks for which AI is becoming well positioned to assist humans. AI is also key to providing robotic assistance on the battlefield, which will enable forces to maintain or expand warfighting capacity without increasing manpower.

Improvements in Cyber Defense

With cyberwarfare as a present and growing military concern—and one that originates in the same digital world as AI—it is natural to expect intersections between the two. These intersections have already begun to manifest as antivirus companies push ever forward in the cat-and-mouse game between attackers and defenders. Historically, one of the ways that antivirus systems have identified malware has been to watch for telltale *static tags*, fixed invisible images that indicate the code is illegitimate.[26] However, it is no longer sufficient to simply use static tags to identify malware, because attackers have discovered ways to generate malware with fewer of those tags. In response, antivirus companies have looked to their large data sets of malware behavior to create AI that can observe software on a system and flag actions that are identified as suspicious. As illustrated by DARPA's Cyber Grand Challenge, there is also growing interest in the potential for machines that can find and patch vulnerabilities in friendly systems or find and attack vulnerabilities in enemy systems, but these applications still cannot perform these tasks at the level of experienced humans.[27]

Improvements in Accuracy and Precision

Machines in general can have greater accuracy and precision than humans. For example, it is possible to use machines to fabricate the electronic transistors that make up computers, despite those transistors being only nanometers across. Machine precision also extends to AI, which can have floating-point precision easily incorporating 32 or 64 bits per number being represented, whereas humans tend to think in rough estimates or round numbers. That is not to say that the number that is being represented with 64-bit precision can be known with certainty, but in principle, the precision is there to accommodate accuracy. Machines can also be more accurate than humans due to certain inherent properties, such as uniformity from machine to machine and uniformity over time, whereas people have more individual differences and are tired or bored.

Labor and Cost Reduction

As is happening throughout the economy, tasks that once required a dedicated person to perform are now progressively being performed by AI or robots. This trend allows a single person to perform quantities of work that would previously have required several people or for some jobs to be automated altogether. The military, a large employer, is no exception and may find ways to reduce staffing levels without sacrificing the services being offered.

Additionally, AI has demonstrated the ability to improve or optimize processes of many different types, which, in turn, leads to cost reductions. With the large number of complex and expensive processes employed by DoD, from logistics to heating and cooling to recruiting, there are plenty of opportunities for AI to improve efficiencies and effect cost savings.

---

[26] Gérard Wagener and Alexandre Dulaunoy, "Torinji: Automated Exploitation Malware Targeting Tor Users," Radu State University of Luxembourg, May 24, 2009.

[27] The 2016 DARPA Cyber Grand Challenge was a competition to create automatic defensive systems capable of reasoning about flaws, formulating patches, and deploying them on a network in real time. See Devin Coldeway, "Carnegie Mellon's Mayhem AI Takes Home $2 Million from DARPA's Cyber Grand Challenge," *TechCrunch*, August 5, 2016.

Improvements in Intelligence, Surveillance, and Reconnaissance

ISR is one of the areas seeing the most current investment in military AI. This trend is likely to continue. The ability to autonomously collect intelligence via drones, from sensors in the terrestrial domain and in space, and even in cyberspace promises to further increase the amount of data being generated. And that volume, velocity, and variety of data will need to be analyzed in part or whole by machines using AI. Some of that analysis will need to be done on ISR platforms deployed in the field, due to bandwidth limitations that make it infeasible to transfer such large quantities of data. Much of the analysis will be done in intelligence processing centers. Wherever it is done, AI will enable dramatic improvements in the quality of intelligence derived from the masses of ISR data collected.

Ability to Operate in Anti-Access/Area-Denial Environments

Potential adversaries have developed capabilities and concepts designed to deny the United States the ability to project force into regions where they may be interested in changing the status quo at the expense of U.S. or allied interests. These capabilities and concepts create what U.S. defense analysts describe as *anti-access/area-denial* (A2/AD) environments, which are increasingly lethal to human operators, platforms, and bases. Autonomous systems will better enable friendly forces to operate in A2/AD environments. Not only will they reduce the numbers of human operators at risk in these environments, but they can also be made smaller, faster, and more agile than inhabited weapons platforms and thus potentially more combat capable. In sum, autonomous weapons and ISR platforms will be able to operate in areas that humans cannot.

Improvements in Deception and Information Operations

The final potential benefit that our expert interviewees suggested has mainly to do with recent trends in international conflict and ML research. It may be possible today or in the near future to have a large number of autonomous agents generating text snippets or short conversations to persuade a target audience to believe a particular narrative of geopolitical or military significance. It is already possible for AI to analyze the large amounts of data that people reveal about themselves online and gain an improved understanding of how to tailor specific messages to increase the likelihood of influencing them.[28] And it is even becoming possible for AI to create false but realistic images, video, and audio of people that could be used maliciously to deceive.[29] It should be noted, though, that respondents to our public survey considered this application to be highly unethical, despite being nonkinetic (discussed further in Chapter 7).

---

[28] Kari Paul, "The Shocking Details You Reveal About Yourself When You 'Like' Things on Facebook," *MarketWatch*, March 25, 2018; Emily Falk and Michael Platt, "What Your Facebook Network Reveals About How You Use Your Brain," *Scientific American*, July 9, 2018.

[29] Dan Robitzski, "AI Can Now Manipulate People's Movements in Fake Videos," *Futurism*, June 6, 2018.

Risks of Artificial Intelligence in Warfare

Although the military applications of AI are expected to yield a wide range of benefits, they also present significant risks. To assess whether employing AI in warfare would be a sound policy choice, one must weigh the expected benefits of these capabilities against the risks they present. Figure 2.3 shows the risks that the experts we interviewed associated with military applications of AI and ranks them by the total number of times each was mentioned during the interviews.

Figure 2.3. Risks of Military Applications of Artificial Intelligence Identified in Structured Interviews

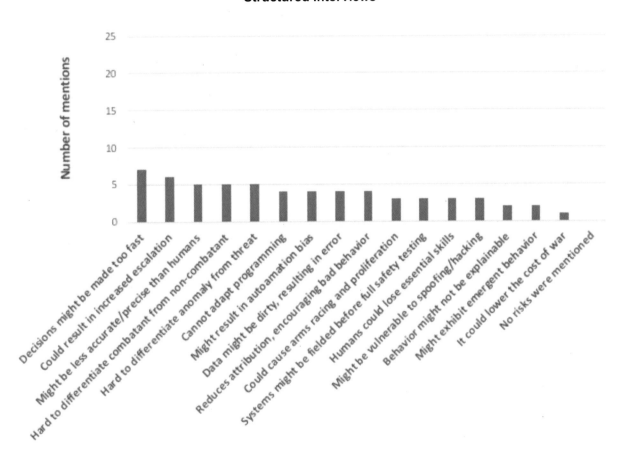

As the figure illustrates, the expert interviews raise significant concerns about military applications of AI. These concerns can be grouped into several broad areas—namely, risks of error, increased risks of war, and risks that military operators and leaders might put too much confidence in these capabilities.

Artificial Intelligence Systems Might Make Dangerous Errors

Although the experts we interviewed named increased speed, accuracy, and precision as potential benefits of military AI, they also expressed concern that these capabilities might make decisions too quickly or that the systems might not be able to adapt to the inevitable complexities of war. As a result, they might not be able to accurately distinguish between combatants and noncombatants or threats and system anomalies, and they may ultimately be less accurate and precise than human operators are. These problems could be magnified if systems are fielded before being adequately tested or if adversaries succeed in spoofing or hacking into them. ML systems could exhibit emergent behavior, acting in dangerous ways.

Artificial Intelligence Could Cause Arms Racing or Escalation

The interviewees also expressed considerable concern that each nation's pursuit of military AI in hopes of gaining a warfighting advantage over potential adversaries could result in proliferation and arms racing. In war, autonomous weapons might not be sufficiently sensitive to political considerations or escalation thresholds. They might attack in places or with levels of intensity that escalate conflicts. The fact that it might be difficult to attribute blame or responsibility to human operators for these acts would complicate matters. This and the possibility that AI could lower the costs of war in terms of human casualties could encourage commanders to take greater risks and act more aggressively, further fueling escalation dynamics.

Military Operators and Leaders Could Put Too Much Trust in Artificial Intelligence

Another concern that our interviewees expressed was that military operators and leaders might put too much trust in their AI systems. They might exhibit "automation bias," relying on the outputs of these AI systems even when they do not seem to make sense. This tendency is intensified in systems in which the algorithmic processing is so complex that their outputs are unexplainable—that is to say, operators cannot easily determine why their systems are giving particular answers or behaving in particular ways.

The Need for a Closer Examination of the Risks of Military Artificial Intelligence

Military applications of AI are advancing at an accelerating pace. Some of the benefits mentioned above are already being realized in systems currently deployed. Other benefits have been demonstrated in controlled applications or lab environments. Still others are anticipated based on projections about what future progress in AI will provide or extrapolations about what military applications might exist for technology that has been demonstrated or is being pursued in the commercial sector.

Yet, for all the potential benefits of military AI, there are significant risks. Some are ethical; others are operational and strategic. It would be ill advised to rush into indiscriminate development, deployment, and employment of these capabilities without examining the risks more closely. We perform that examination in the next chapter.

3. Risks of Military Artificial Intelligence: Ethical, Operational, and Strategic

This chapter provides an overview of ethical concerns associated with AI in military applications, develops a taxonomy of ethical and other risks of military AI, and concludes by considering perspectives on how to mitigate risks.

Stakeholder Concerns About Military Artificial Intelligence

Contemporary discussions of AI are often influenced by depictions of AI systems and robots in popular narrative. Even long before the beginning of modern robotics, there were stories of human-created anthropomorphic beings that rebel against their creators and run.[1] In the mid-1900s, with the rise of robotics and computing systems, science fiction writers such as Stanislaw Lem and Isaac Asimov explored the conceptual and technical challenges associated with communicating with robots and ensuring their safety and reliability. In the later twentieth century, films such as *The Terminator* and *The Matrix* spread visceral images of extreme risks to humanity stemming from self-aware systems exhibiting artificial general intelligence (AGI). In these narratives, human-created systems surpass human intelligence and capabilities and pursue objectives that are not aligned with human interests. Consequently, they present an existential risk to all of humanity. Fictional narratives continue to resonate and shape people's perception of AI and have been used to advocate for policy prescriptions. For instance, the short "Slaughterbots" video, produced by the Future of Life Institute, depicts the malicious use of highly mobile armed systems that can autonomously identify and attack specified targets. This video, which implores its viewers to "stop autonomous weapons," has had over 2.5 million views.[2]

Whether or not these popular narratives accurately depict future possibilities, they have cast frightening and psychologically resonant scenarios that inform public discussion of AI. But it is not only fiction writers and the film industry who have expressed concern about risks; AI technologists and technology companies have also recently raised alarms. In addition, international efforts to limit the development or use of AI systems in military applications have begun to coalesce. Indeed, a wide range of actors has expressed concerns about the risks of military AI.

Technologists

Some of the most vocal proclaimers of the risks of AI are prominent scientists and technologists involved in AI R&D. They focus less on the long-term, existential risks associated

---

[1] For instance, consider traditional Jewish stories of the Golem or Mary Shelley's *Frankenstein*.

[2] StratoEnergetics, Buenos Aires Event, "Slaughterbots," YouTube, November 12, 2017.

with AGI or "Skynet,"[3] than on the shorter-term risks related to the use of systems that might plausibly be developed over the coming years. The most noteworthy example is the 2015 "Open Letter from AI & Robotics Researchers," which expresses concern about an AI arms race and the possible proliferation of lethal AI systems to terrorists or dictators.[4] The letter, signed by almost 4,000 researchers to date, including prominent technologists such as Stuart Russell, the late Stephen Hawking, and Elon Musk, claims that "most AI researchers have no interest in building AI weapons." It warns that the development of military AI systems could produce a public backlash that constrains the beneficial applications of AI. To ensure that AI is able to offer purported social benefits, the letter advocates for a "ban on offensive autonomous weapons beyond meaningful human control."

This chapter will later return to the proliferation challenges and risk-mitigation strategies, such as the possibility of an international ban and the concept of meaningful human control. The point to emphasize here is that some of the researchers closest to and most influential in AI development have expressed alarm. Their efforts extend beyond signing open letters. Many are directing their research efforts to activities that increase the reliability, explicability, and safety of different AI systems.

Industry

A related set of ethical concerns surrounding military AI has been raised by the private sector and, specifically, by employees at certain technology companies. One early example was Clearpath Robotics, a leader in unmanned systems development, which in 2014 issued a statement noting that "the negative implications of [autonomous weapon systems] far outweigh any benefits" and that the company "would not manufacture weaponized robots that remove humans from the loop."[5]

More recently, a letter from the staff of Google, reportedly signed by over 4,000 employees, asked the chief executive officer to "draft, publicize and enforce a clear policy stating that neither Google nor its contractors will ever build warfare technology."[6] The letter particularly raised alarm about Google's work with DoD on Project Maven to build a customized AI engine to detect objects in imagery collected by drones.[7] In response to this pressure, Google decided not

---

[3] "Skynet" is the fictional AGI system depicted in the *Terminator* movie franchise. It is an artificial group consciousness that controls the Terminator robots and is the real antagonist in these movies.

[4] Future of Life Institute, "Autonomous Weapons: An Open Letter from AI & Robotics Researchers," July 28, 2015.

[5] Meghan Hennessey, "Clearpath Robotics Takes Stance Against 'Killer Robots,'" August 13, 2014.

[6] Scott Shane and Daisuke Wakabayashi, "The Business of War: Google Employees Protest Work for the Pentagon," April 4, 2018.

[7] Project Maven, also known as the Algorithmic Warfare Cross-Function Team, was launched in April 2017. It is an effort to develop and integrate computer-vision algorithms and other AI capabilities to process ISR data and, in particular, the large volume of full-motion video data that DoD collects every day in support of counterinsurgency and counterterrorism operations. See Adam Frisk, "What Is Project Maven? The Pentagon AI Project Google Employees Want Out Of," *Global News*, April 5, 2018.

to renew the Project Maven contract with DoD.[8] Although the image-recognition technologies involved in Project Maven are not "autonomous weapon systems" of the sort envisioned by the signers of the "Open Letter from AI & Robotics Researchers" or Clearpath Robotics, Google faced significant bottom-up pressure from its staff about this specific military application.[9]

Google also published a set of AI guiding principles—the first of its kind for a major technology company. These principles articulate an ethical standard for when it will develop AI—namely when it "believe[s] that the overall likely benefits substantially exceed the foreseeable risks and downsides."[10] The AI principles state that Google will continue working with the U.S. government and military but contends that Google will not design or deploy AI for a variety of specific applications. These include "weapons or other technologies whose principal purpose or implementation is to cause or directly facilitate injury to people"; "technologies that gather or use information for surveillance violating internationally accepted norms"; or "technologies whose purpose contravenes widely accepted principles of international law and human rights."[11]

It is not clear how these new principles will be put into practice or what future opportunities with the military the principles preclude. For instance, the principles do not seem to apply to weapons whose main purpose is to damage military objects or to cyber capabilities whose main purpose is not a kinetic effect. Indeed, Google has pursued other military contracts—for instance, it has marketed its cloud capabilities to assist Special Operations Forces with "sensitive site exploitation"—and it is uncertain whether it will continue to pursue these types of opportunities.[12] However, the statement of concern and prohibition on a potentially wide swath of military applications from a technology leader will shape the context and contours of at least some forms of public-private partnership.

There are other examples of industry limiting its role with the military due to ethical concerns. DeepMind, one of the most advanced centers of ML and AI research and an autonomous component of Google's parent company, Alphabet, has sought to prohibit military applications of its research. When it was acquired by Alphabet, it inserted a provision in the acquisition agreement that its technologies would not be used for military purposes.

Many other American companies, both within the traditional defense industrial base and newer technology companies, such as Palantir, develop AI technologies for the U.S. military without the same apparent apprehension. There are many questions as to whether Google's

---

[8] Daisuke Wakabayashi and Scott Shane, "Google Will Not Renew Pentagon Contract That Upset Employees," *New York Times*, June 1, 2018.

[9] In Chapter 7, we describe findings from our survey of the public's view of military programs such as Project Maven.

[10] Sundar Pichai, "AI at Google: Our Principles," *AI*, June 7, 2018.

[11] Pichai, 2018.

[12] Patrick Tucker, "Here's How Google Pitched AI Tools to Special Operators Last Month," *Defense One*, June 10, 2018.

AI principles portend a trend that other major U.S. technology companies, such as Amazon and Microsoft, will adopt.[13] There are also questions about whether employees of Chinese or other international firms will raise ethical concerns about developing technologies for military application. Indeed, Google has also opened an AI research center in Beijing.[14] However, with an AI-savvy workforce continuing to be in high demand, companies such as Google have sought to accommodate and address their employees' ethical concerns.

International Advocacy

Other groups and individuals, including nongovernmental organizations, faith leaders, and academics, have expressed their support of international advocacy efforts to limit the development or use of autonomous weapons. The two most noteworthy groups, the International Campaign for Robot Arms Control, begun in 2009, and the Campaign to Stop Killer Robots, a coalition launched in 2013, have published a series of reports and organized events to mobilize nation-states, industry, and the public. Some members of these groups were also active in other arms control and humanitarian activities, such as the efforts to develop the Ottawa Mine Ban Treaty and the Treaty on the Non-Proliferation of Nuclear Weapons. Their efforts regarding military AI are directed at promoting "a comprehensive, pre-emptive prohibition on the development, production and use of fully autonomous weapons—weapons that operate on their own without human intervention" through an international treaty, national laws and other measures.[15] These groups have detailed the ethical and legal risks they believe arise with military AI systems, which we will explore below in the taxonomy of risks.

Religious and other cultural leaders have also spoken about the ethics of AI, and some have supported a ban or other regulatory control. In 2014, an interfaith declaration supporting a ban on autonomous weapons was signed by dozens of religious leaders, including Archbishop Desmond Tutu.[16] The Dalai Lama and other Nobel Peace laureates have also supported declarations calling for a preemptive ban on lethal autonomous weapons, which they see as a "new form of inhumane warfare."[17] And the Holy See—the sovereign entity of the Catholic

---

[13] Amazon employees have expressed concerns about working with Immigration and Customs Enforcement (ICE) on facial-recognition technologies. See Hamza Shaban, "Amazon Employees Demand Company Cut Ties With ICE," *Washington Post*, June 22, 2018. Note also that Amazon Chief Executive Officer (CEO) Jeff Bezos seemed to support a ban on fully autonomous weapons in Forum on Leadership, "Closing Conversation with Jeff Bezos, Co-Presented with SMU," Dallas, Tex.: George W. Bush Presidential Center, April 20, 2018.

[14] Jonathan Vanian, "Google Plans Big AI Push in Asia," *Fortune*, December 13, 2017.

[15] "Campaign to Stop Killer Robots: The Solution," *ST THOMAS AQUINAS VERSUS NASA*, March 10, 2018.

[16] "Religious Leaders Call for a Ban on Killer Robots," *PAX*, December 12, 2014.

[17] "World Summit of Nobel Peace Laureates: Final Declaration," Pressenza, December 14, 2014.

Church—produced a paper on autonomous weapons in which it states that "it is fundamentally immoral to utilize a weapon the behavior of which we cannot completely control."[18]

It is too early to assess how successful these efforts will be at marshaling public support. Unlike landmine and chemical weapon bans, where abhorrent images of humanitarian harms drew public attention and support for regulation, there are not actual examples of fully autonomous weapon systems in use and involved in atrocities. That said, narratives from science fiction have already produced compelling images that have galvanized fear, and advocacy groups regularly point to the risks contained in such narratives in support of their efforts.

United Nations Convention Discussions

The United Nations Convention on Certain Conventional Weapons (UN CCW) is an international treaty that entered into force in 1983 and is intended to prohibit or restrict the use of weapons that are "excessively injurious or have indiscriminate effects."[19] The UN CCW operates as a framework, or chapeau convention, that provides a forum to negotiate additional protocols to prohibit or restrict specific weapon systems. There are currently five protocols of the UN CCW, including protocols restricting or prohibiting the use of nondetectable fragments, landmines, incendiary weapons, blinding laser weapons, and explosive remnants of war.

In 2014, the UN CCW held the first multilateral meeting focused on challenges stemming from LAWS; this was followed by other informal meetings in 2015 and 2016. In 2017 and 2018, the UN CCW convened in a more formalized structure known as a Group of Governmental Experts (GGE). The GGE meetings are tasked to consider the definition of LAWS, the role of the human in using lethal force, and possible options for addressing humanitarian and security challenges.[20] One option up for discussions is whether there is a need for a new formal additional protocol that prohibits the use of fully autonomous weapons. The Campaign to Stop Killer Robots claims that 26 countries have supported a ban on LAWS under the CCW process. At the August 2018 GGE meeting, Austria, Brazil, and Chile pushed to move discussions toward treaty negotiations. However, the United States, Russia, and the United Kingdom stated their opposition to new formal legal instruments. Many governments, including the United States, Russia, and China, have also submitted official statements outlining their views on LAWS.[21] Technologists,

---

[18] "Elements Supporting the Prohibition of Lethal Autonomous Weapons Systems," working paper submitted by the Holy See to the Group of Governmental Experts of the High Contracting Parties to the Convention on Prohibitions or Restrictions on the Use of Certain Conventional Weapons Which May Be Deemed to Be Excessively Injurious or to Have Indiscriminate Effects, April 7, 2016, p. 8.

[19] United Nations Office at Geneva, *Convention on Certain Conventional Weapons*, United Nations Publication, 2014, p. 1.

[20] For information on the 2018 GGE meeting, including the agenda, list of participants, and links to the working papers presented there, see UNOG, "Group of Governmental Experts on Lethal Autonomous Weapons Systems (LAWS)," 2018a.

[21] We discuss the policy and ethical dimensions of the UN CCW national submissions in Chapters 4–6.

academic experts, and others have participated in the meetings to offer perspectives, and international groups have organized side events advocating bans or other restrictions.

As this overview indicates, a range of stakeholders from across sectors have expressed concern about military AI. Too often, however, such concerns are expressed in overly general, emotional-laden terms, raising ethical concerns about "killer robots" that are imprecise. The taxonomy of AI risks below seeks to more systematically describe and categorize the range of risks that these stakeholders have identified.

Taxonomy of Artificial Intelligence Risks

War is an inherently risky activity. Belligerents use deadly force in efforts to obtain military and political objectives, and the effects of interactions between opposing military forces are highly uncertain, not only for the combatants but for noncombatants as well. However, there are many kinds of risk, and some weapons and uses of force are riskier than others. This section presents a taxonomy of risks associated with military AI. We have identified these risks based on a review of the relevant literature, including works produced by the stakeholders described above, and interviews with select experts. As part of the expert interviews, we developed a series of vignettes depicting applications of AI systems in military contexts and asked the interviewees to comment on the plausibility of each scenario and the risks associated with employing the AI systems depicted in it.[22] Figure 3.1 presents our taxonomy of AI risk.[23]

As the figure indicates, we have organized the risks of military AI into three categories: ethical and legal, operational, and strategic. Many risks could fall into several different categories; this report organizes them in a way that we believe is clearest and most consistent. Per the scope of this report, we will limit our discussion of risks to those associated with military AI systems already in development or with a real possibility of deployment over the next 10–15 years. Given the state of AI development and likely technological trajectory, the 10–15-year time frame will not likely lead to the creation of an artificial general superintelligence of the sort exemplified by "Skynet" in the *Terminator* films. Thus, the existential risk of a superintelligent robot intent

---

[22] The vignettes are provided in Appendix A.

[23] The focus of this report is on the ethical implications of military applications of AI. We grant that many of the risks associated with military AI, such as risks we have categorized as operational, might not be thought by all people as ethical in nature. However, the line between the right or best actions from an ethical perspective or from strategic, operational, or some other perspective is not sharply defined and might depend on one's first-order ethical theory. Ethics is a broad concept with a myriad of interpretations. Ethicists have articulated several prominent first-order normative ethical theories, most notably utilitarianism and deontology, that can be applied to the issue of the ethical permissibility of military AI. However, this is a fragmented space where people's ethical outlooks differ and are not always clear, consistent, or comprehensive. And even when people agree about basic ethical principles, or share a single normative theory, they might interpret these principles differently and reach different conclusions about how they should be applied. In the chapters on military AI developments in the United States, Russia, and China, we will discuss how those respective countries might interpret and respond to these risks.

Figure 3.1. Taxonomy of Artificial Intelligence Risk

Ethical and Legal	Operational	Strategic
Law of Armed Conflict Accountability and Moral Responsibility Human Dignity Human Rights and Privacy	Trust and Reliability Hacking, Data Poisoning, and Adversarial Attacks Accidents and Emergent Risks	Thresholds Escalation Management Proliferation Strategic Stability

on eliminating all humans will not be discussed here. Finally, this taxonomy of risks is intended to provide an overview, not to be exhaustive. We shall address each of these risks in the following sections.

Ethical and Legal

This section discusses ethical and legal considerations related to military AI. Ethical considerations sometimes undergird legal measures, but there might also be ethical considerations not codified into current law. On the other hand, legal structures provide formal constraints that shape behavior, usually provide a system to administer consequences for violations, and thereby provide an additional reason for taking or refraining from action.

Law of Armed Conflict

LOAC, also known as International Humanitarian Law (IHL), is intended to regulate the conduct of hostilities and minimize humanitarian harms to civilians.[24] LOAC is codified in formal treaties such as the Four Geneva Conventions, the Additional Protocols, and weapon-specific treaties such as the Ottawa Landmine Ban. It is also based in customary state practice supported by *opinio juris*.[25] Some of the key principles of LOAC broadly recognized by national governments are

1. Distinction: Belligerents must distinguish between civilians and combatants and direct military operations only against military objectives;
2. Proportionality: Belligerents are prohibited from actions that cause excessive harm that is disproportionate to the military objective; and

---

[24] This report does not offer the official view of DoD's approach to LOAC.

[25] *Opinio juris* is the opinion or belief that a specific action is legally required.

3. Precaution: Belligerents must take steps to caution against harming civilians.

Another key principle of international law relevant to military AI includes the requirement of a legal review of weapons to ensure LOAC compliance.[26] Finally, LOAC is generally interpreted to include the so-called Martens Clause, which is a principle in Additional Protocol 1 that refers to protections that extend beyond codified law and are based on the "principles of humanity and the dictates of public conscience."[27]

Some critics have argued that fully autonomous weapon systems (defined as weapons that select and engage targets without human authorization) are incapable of complying with LOAC's principle of distinction and proportionality.[28] They argue that an AI system does not have the capacity to understand and assess the subtle indications such as body language that would allow the distinction "between a fearful civilian and a threatening enemy."[29] Distinguishing combatants from noncombatants might be especially challenging in the context of asymmetric conflict in urban settings, where combatants do not always wear uniforms or other insignia. Especially in these settings, only a human and not an autonomous weapon could comply with the principle of distinction. The use of AI-enabled cyber capabilities also raises questions about the principle of distinction, since cyber capabilities usually need to travel over civilian networks and are often targeted against civilian-owned or -operated systems.[30] Critics have also argued that autonomous weapon systems cannot satisfy the principle of proportionality, because it requires a subjective case-by-case assessment of the harm of possible collateral effects weighed against the importance of the military objective. Making a judgment of proportionality, critics argue, "requires more than a balancing of quantitative data"; it entails an evaluative, qualitative, and ethical assessment by a human weighing and comparing complex values.[31] The upshot of these arguments is that existing international law prohibits the use of autonomous weapons.

Responses to these arguments have noted that there is a conceptual distinction between the illegality of autonomous weapons, per se, and specific types or uses of autonomous weapons. As

---

[26] This is codified in Art 36 of Additional Protocol 1: "In the study, development, acquisition or adoption of a new weapon, means or method of warfare, a High Contacting Party is under an obligation to determine whether its employment would, in some or all circumstances, be prohibited by this Protocol or by any other rule of international law." *Protocols Additional to the Geneva Conventions of August 12, 1949*, relating to the Protection of Victims of International Armed Conflicts (Protocol I), Article 36, June 8, 1977, p. 30. Although the United States is not a party to Additional Protocol 1, it has stated that it has a long-standing policy requiring legal review of weapons.

[27] Rupert Ticehurst, "The Martens Clause and the Laws of Armed Conflict," *International Review of the Red Cross*, No. 317, April 30, 1997.

[28] See, for example, HRW and International Human Rights Clinic, *Losing Humanity: The Case Against Killer Robots*, 2012; HRW and International Human Rights Clinic, *Making the Case: The Dangers of Killer Robots and the Need for a Preemptive Ban*, 2016.

[29] HRW and International Human Rights Clinic, 2012, p. 4.

[30] See, for instance, Michael N. Schmitt, ed., *Tallinn Manual 2.0 on the International Law Applicable to Cyber Operations*, Cambridge, UK: Cambridge University Press, 2017.

[31] HRW and International Human Rights Clinic, 2012, p. 39.

some scholars explain, "Individual systems could be developed that would violate these norms, but autonomous weapon systems are not prohibited on this basis *as a category*."[32] Indeed, nearly all weapons could be used in ways that violate LOAC, so the important consideration is how they are used. These responses also argue that there will likely be continuous improvements to autonomous weapons, some of which will better enable these systems to comply with LOAC. They also point to cases in which there is low risk of civilian harm from the use of autonomous weapons—for instance, on the open oceans, under the sea, or with regard to isolated military targets. In these cases, they argue, the systems do not have to make difficult qualitative evaluations of distinction and proportionality and therefore could be fully LOAC compliant.

This is not the place to arbitrate this debate, but it is worth noting that DoD acknowledges the importance of LOAC compliance. DoD's *Law of War Manual* states that although there is no law prohibiting the use of autonomy in weapon systems, the regular principles of LOAC do apply to their use.[33] Indeed, the *Law of War Manual* notes that in many cases "the use of autonomy could enhance the way law of war principles are implemented in military operations," for instance by improving the military's ability to execute precise attacks with limited collateral effects.[34] It contends, however, that "the law of war rules on conducting attacks (such as the rules relating to discrimination and proportionality) impose obligations on persons. These rules do not impose obligations on the weapons themselves."[35] Some have argued that this statement precludes the use of fully autonomous weapons outside human control since those systems might be designed to operate without a human making judgments about distinction, proportionality, and precaution for each attack.[36] This issue will be discussed further below.

Accountability

A related ethical risk is that the use of autonomous weapons creates an accountability gap. Accountability serves several purposes. First, it acts as a deterrent against conducting harmful actions through credible threats of punitive measures against the responsible actors. Second, it ensures that a specific actor has responsibility for taking steps to ensure compliance with the legal factors relevant to the action. Third, accountability acts as an important moral concept that designates moral responsibility for an action, including determining reasonable moral retribution,

---

[32] Michael Schmitt, "Autonomous Weapon Systems and International Humanitarian Law: A Reply to Critics," *Harvard Law School National Security Journal*, February 5, 2013.

[33] DoD, *Law of War Manual*, June 2015, p. 330. This report does not offer the official view of DoD's approach to LOAC.

[34] This is also discussed in the August 2018 U.S. submission to the UN CCW GGE, "Human-Machine Interaction in the Development, Deployment and Use of Emerging Technologies in the Area of Lethal Autonomous Weapon Systems," August 2018.

[35] DoD, 2015, p. 330.

[36] See Heather Roff, "Meaningful Human Control or Appropriate Human Judgment? The Necessary Limits of Autonomous Weapons," Briefing Paper for Delegates at the Review Conference of the Convention of Certain Conventional Weapons, Geneva, December 2016.

appropriate feelings of moral emotions such as shame and guilt, and responsibility for making moral amends or redress. Thus, accountability serves an important deterrent, legal, and moral role not just in war but throughout social life.

Critics have contended that fully autonomous weapons will make decisions without enabling any proper locus of accountability, thereby negating its deterrent, legal, and moral value.[37] These critics argue that it does not make conceptual sense to hold a nonhuman weapon system itself as having some legal or moral responsibility for its actions. The fact that the system makes its own decisions seems to create distance from and mitigate the responsibility of the military operators or commanders using the system. The system's developers, programmers, or testers similarly cannot be reasonably said to be accountable for uses of the system that they had no part in choosing. For each of these actors, the use of autonomous systems might give rise to a "moral buffer" between humans and the systems' actions.[38] Other types of legal accountability, such as liability regimes for negligence familiar in domestic criminal or civil law, are not available within LOAC. Furthermore, for at least some AI military applications such as cyber operations, the challenge of attribution also poses a potential accountability challenge. If it is not possible to attribute the source of an attack, then it is not possible to hold the perpetrator accountable. This attribution challenge also enables the possibility of false-flag operations.[39] For these reasons, advocates argue, there are no proper focal points of responsibility for autonomous weapon systems in war, and without a focal point of responsibility, there is an accountability gap.[40]

Commentators have responded that this is not a genuine risk and that there might be clear cases of who is accountable for decisions in the military chain of command.[41] As mentioned above, the DoD *Law of War Manual* holds that "it is persons who must comply with the laws of war," and thus there is a locus of responsibility—namely, the person who uses the weapons.[42] Indeed, the manual states that the person using autonomous weapons "must refrain from using that weapon where it is expected to result in incidental harm that is excessive."[43] It even notes that the obligations to take precaution "may be more significant when the person uses weapon systems with more autonomous functions."[44] This might help specify that according to DoD,

---

[37] For one articulation of this view, see HRW and International Human Rights Clinic, *Mind the Gap: The Lack of Accountability for Killer Robots*, April 2015.

[38] The phrase "moral buffer" comes from M. L. Cummings, "Creating Moral Buffers in Weapon Control Interface Design," *IEEE Technology and Society Magazine*, Fall 2004.

[39] A false-flag operation is one designed to create the appearance of a particular party, group, or nation being responsible for some activity, disguising the actual source of responsibility.

[40] Latiff, 2017, p. 48.

[41] See Charles Dunlap, "Accountability and Autonomous Weapons: Much Ado About Nothing?" *Temple International and Comparative Law Journal*, Vol. 30, 2016, pp. 63–76.

[42] DoD, 2015, p. 330.

[43] DoD, 2015, p. 330.

[44] DoD, 2015, p. 330.

there is a person accountable when using autonomous weapons. However, critics contend that it is not clear who is appropriately described as "using" an autonomous weapon, and thus there is ample room to pass on responsibility to others or, worse yet, to no one. In addition, the high risk of accidents and emergent effects with AI systems makes it hard to assess whether a harmful outcome can be appropriately deemed to be the responsibility of the operator, developer, or tester or is rather just an unpredictable development with no one ultimately at fault.

Arguments from Human Dignity

Dignity is a key moral concept in a variety of ethical and religious traditions. The Universal Declaration of Human Rights begins by stating that "recognition of the inherent dignity and of the equal and inalienable rights of all members of the human family is the foundation of freedom, justice and peace."[45] Some advocates have argued that the inherent dignity of humans entails that it is always wrong for an autonomous weapon or other machine to take a human life. This is not a matter of the rightfulness or wrongfulness of killing any particular person in conflict, but the process or method of how that person is killed. According to this view, only humans capable of making moral judgments and possessed of their own inherent dignity are morally able to take the lives of others. Nonhuman systems do not have the necessary moral qualities to justify their actions in ways that respect victims and thus should not make decisions with such significant ethical implications. This view has been articulated by the former UN Special Rapporteur on Extrajudicial, Summary, or Arbitrary Executions Christof Heyns, who has argued that "death by algorithm means people are treated as interchangeable entities.... A machine, bloodless and without morality or mortality, cannot fathom the significance of the killing or maiming of a human being."[46] Vice Chairman of the Joint Chiefs General Paul Selva has made similar points, noting in Senate testimony that he does not think "it is reasonable for us to put robots in charge of whether or not we take a human life."[47]

This debate is closely related to discussions surrounding the moral importance of human emotions in the conduct of warfare. On the one hand, human emotions, such as fear or anger, might lead a human to make unwise decisions or mistakes that result in serious harms, including harms to innocent civilians. On the other hand, Human Rights Watch (HRW) and others have argued that "emotions should . . . be viewed as central to restraint in armed conflict rather than as irrational influences and obstacles to reason" and that "human emotions . . . provide one of the

---

[45] United Nations, *The Universal Declaration of Human Rights*, December 10, 1948.

[46] Christof Heyns, "Autonomous Weapons Systems and Human Rights Law," Presentation Made at the Informal Expert Meeting Organized by the State Parties to the Convention on Certain Conventional Weapons, May 13–16, 2014, Geneva, Switzerland.

[47] U.S. Congress, Senate Committee on Armed Services, "Hearing to Consider the Nomination of General Paul J. Selva, USAF, for Reappointment to the Grade of General and Reappointment to Be Vice Chairman of the Joint Chiefs of Staff," Washington, D.C., July 18, 2017. Other senior U.S. military officers have also expressed these views. For instance, see Latiff, 2017, p. 47.

best safeguards against killing civilians."[48] Emotions enable warfighters to have the sense of compassion and respect for human life that is crucial to exercising caution necessary in the fog of war. According to this argument, systems that do not possess emotions lack a basic human feature essential to any just warfare.

Whether or not human dignity is threatened by AI is a complex ethical matter that cannot be resolved here. However, the view that AI, and more specifically autonomous weapons, pose risks to human dignity resonates broadly among the public, as does the view that emotions play an important role in just war.

Human Rights, Privacy, and Truth Decay

A final set of ethical risks relates to the many potential threats to human rights and individual privacy from AI. AI-enabled systems, such as big data analysis, persistent ISR, face recognition, and cyber capabilities, might enable autocrats to surveil their population, target dissidents, censor content, or otherwise infringe on basic human rights.[49] Military investment and development of AI applications will potentially further increase the proliferation of tools that can be used maliciously. Networked devices, commonly called the "internet of things," mobile phones, and other network connections provide opportunities for monitoring and surveillance. Information operations that spread false information and take advantage of cognitive and social biases might be thought to produce other social harms such as the diminished importance of objective facts and data—a phenomenon known as "Truth Decay."[50] Russia has already used online social media bots that pretend to be human to shape American and other countries' discourse and more generally to sow discord. The ability to develop capabilities that create credible fake videos adds additional risks to the foundation of common beliefs that are arguably essential for democratic social stability. In addition, ML systems might produce outputs that unfairly discriminate against minorities or other groups due to biased or nonrepresentative training data. Some critics have further contended that AI is a potentially revolutionary enabler of radical social change, with the possibility of a myriad of other human rights and social implications, and they argue that increased military investment will hasten the development of technologies with these undesirable effects.

Operational

Operational risks refer to risks associated with the intended functioning of AI in military applications. These are risks involving ways in which the use of military AI might fail in unintended or unanticipated ways. Depending on one's theoretical or philosophical outlook,

---

[48] HRW and International Human Rights Clinic, 2012, pp. 27, 37.

[49] Latiff, 2017, p. 29.

[50] See Jennifer Kavanagh and Michael D. Rich, *Truth Decay: An Initial Exploration of the Diminishing Role of Facts and Analysis in American Life*, Santa Monica, Calif.: RAND Corporation, RR-2314-RC, 2018.

these may or may not be considered ethical in nature. Regardless, these risks have been noted as significant concerns by a variety of AI technologists and other experts.

Trust and Reliability

Risks of trust and reliability concern those associated with both not trusting systems sufficiently and overtrusting them. Regarding the former, there are several hurdles to developing trust in AI-enabled systems. First, there is what is sometimes referred to as the "black box" problem, or the problem that AI systems, such as those enabled by deep neural nets, reach conclusions and produce outputs in ways not evident or easily explained to humans. The issue is that the nature of complex algorithms does not allow one to easily work back through a system's processing to understand why it reached the output it did. If human operators do not understand how the system they are using works, they might not develop the trust in the system necessary to use it appropriately, thereby precluding potential benefits of the system or misusing it in potentially harmful ways. A second and related factor is the challenge of testing and evaluating AI-enabled systems. Many of these systems are designed to be deployed in complex and unstructured environments. Testing in closed ranges or laboratory settings might not sufficiently ensure that these systems will behave as intended once they are deployed. The two issues of AI explainability and testing are mutually reinforcing and might create a trust deficit, as indicated in the following remarks by General Selva:

> The entire enterprise of software engineers who are trying to build this artificial intelligence, deep learning space have not built the piece of software that can actually tell you what it's learned. And I think that's one of the milestones we're going to have to cross before we can actually get into a high-confidence area where we can say that technology is actually going to do what we want it to do, because not only can we physically test it; we can intellectually test it.[51]

These remarks suggest the difficulty of trusting systems that cannot explain their decisionmaking in ways understandable to operators, which creates risks of misunderstanding or misusing AI systems.[52]

The challenge of trust regarding AI also extends the other way such that operators or commanders may have excessive trust in AI systems, even when those systems make errors. Studies in aviation and other applications have shown problems stemming from "automation bias," which "occurs when a human decision maker disregards or does not search for contradictory information in light of a computer-generated solution which is accepted as

---

[51] Joint Chiefs of Staff, "Gen. Selva's Q&A Session at the Brookings Institution," Brookings Foreign Policy Program, January 2016.

[52] There is work on so-called explainable AI to help grapple with these problems. For instance, DARPA's Explainable AI program seeks to develop techniques to produce more explainable models that enable increased human understanding and trust in the system. See David Gunning, "Explainable Artificial Intelligence (XAI)," DARPA, n.d.

correct."[53] This overconfidence in the system can have far-reaching implications. For instance, a well-known study on fratricide incidents during Operation Iraqi Freedom found that a major factor in these incidents was operators "reacting quickly, engaging early, and trusting the system without question."[54] As systems become more regularly used in military contexts, there is increased risk of automation bias and its attendant harms.

Hacking, Data-Poisoning, and Adversarial Attacks

Another oft-noted risk of AI military systems stems from vulnerabilities of these systems to malicious actors. In light of the multitude of cyber incidents occurring in military and civilian networks alike, it is clear that all network-enabled technologies are vulnerable to hacking, especially by a determined and well-resourced adversary. Even closed or air-gapped networks are vulnerable to supply-chain attacks or other means of gaining access for malicious purposes. ML systems that learn from training data are vulnerable to other types of attacks, including so-called data-poisoning attacks, in which the training data is manipulated or spoofed in order to influence the intended functioning of the system. AI-enabled systems can also fail due to adversarial attacks intentionally designed to trick or fool algorithms into making a mistake.[55] These examples demonstrate that even simple systems can be fooled in unanticipated ways and sometimes with potentially severe consequences.[56]

Of course, AI-enabled systems might also help defend against hacking, data-poisoning, or other types of malicious attacks. It is unclear whether offense or defense will ultimately have the advantage in these applications. Given this uncertainty, and the breadth of research demonstrating the vulnerabilities of current AI systems, stakeholders have noted that these risks need to be taken seriously in the context of military AI.

Accidents and Emergent Risks

Accidents are inevitable even in simple automated systems. Studies have shown that as systems become more complex and "tightly coupled" with other systems—creating "systems of systems"—the risks of unintended accidents increase.[57] There are myriad of examples of

---

[53] See definition and references to studies in M. L. Cummings, "Automation Bias in Intelligent Time Critical Decision Support Systems," AIAA 1st Intelligent Systems Technical Conference, September 22, 2004.

[54] John K. Hawley, "Looking Back at 20 Years of MANPRINT on Patriot: Observations and Lessons," Army Research Laboratory, ARL-SR-0158, September 2007.

[55] See overview and references in OpenAI, "Attacking Machine Learning with Adversarial Examples," February 24, 2017.

[56] For example, see research using physical adversarial examples that trick AI systems in self-driving vehicles into not perceiving a stop sign. Ivan Evtimov, Kevin Eykholt, Earlence Fernandes, and Bo Li, "Physical Adversarial Examples Against Deep Neural Networks," Berkeley Artificial Intelligence Research, December 2017.

[57] For instance, see United Nations Institute for Disarmament Research (UNIDIR), "Safety, Unintentional Risk and Accidents in the Weaponization of Increasingly Autonomous Technologies," *UNIDIR Resources*, No. 5, 2016; Paul Scharre, "Autonomous Weapons and Operational Risk," Center for a New American Security, February 2016.

complex system failures, including the Chernobyl meltdown and the *Challenger* disaster. The development of AI, ML, and autonomous systems exacerbates these risks since they increase complexity and speed, thereby making it more challenging for human operators to predict where problems will occur and to supervise and monitor systems in use.[58] The 2010 stock market "Flash Crash" is an often-cited example of autonomous systems interacting at high speeds and with negative unpredicted results—in this case, a significant drop of the Dow Jones in minutes.[59] Interaction risks such as these are not evident when analyzing a single system; rather, they occur only in the system's interactions in complex environments. Other examples of AI system accidents include the apparent unintended release of the Stuxnet worm onto the open internet and other cases of self-replicating malware such as in the WannaCry and NotPetya attacks. These examples show the hard-to-control nature of systems with autonomous capabilities.

Former Secretary of the Navy Richard Danzig has written about the risks of complexity and accidents stemming from increasingly autonomous systems, arguing that "progress toward our primary goal, superiority, should be expected to increase rather than reduce collateral risks of loss of control."[60] He notes several features of AI systems in military applications that "make them especially prone to human error [and] emergent effects," including the secrecy of systems, the unpredictability of military environments, the mismatch between experts' skills and their military assignments, the interdependencies of systems, the pressure to rapidly deploy, and the pressure from international competitors. Due to these factors, AI in military applications might pose higher risks of accidents and other failures than in other AI applications.

Strategic

Strategic risks are those that give rise to significant challenges to national-level objectives. The supposed benefits of military AI have encouraged investment not just by the United States but also by China, Russia, and other actors. These developments give rise to concerns about the stability of the international order.

Thresholds

As discussed in Chapter 2, one of the advantages of military AI, such as advantages stemming from unmanned aerial vehicles (UAVs), is that these systems can be used in "dull, dirty, and dangerous missions."[61] The deployment of autonomous systems thereby potentially lowers the risks of harm to human military personnel. However, critics have noted that this

---

[58] For a discussion on a range of accident risks in ML systems, see Dario Amodei, Chris Olay, Jacob Steinhardt, Paul Christiano, John Schulman, and Dan Mané, "Concrete Problems in AI Safety," July 25, 2016.

[59] Further analysis of this flash crash is in Scharre, 2018, pp. 199–207.

[60] See Richard Danzig, "Technology Roulette: Managing Loss of Control as Many Militaries Pursue Technological Superiority," Center for a New American Security, June 2018, p. 2.

[61] Office of the Secretary of Defense, *Unmanned Systems Roadmap (2007–2032)*, Washington, D.C.: Department of Defense, December 2007, p. 19.

feature of autonomous systems also creates the risk that leaders will resort to using armed autonomous systems instead of pursuing nonmilitary options. In this way, the threshold for military action will be lower, and with more military action, there will be greater costs to innocent civilians. HRW, for instance, argues that the lower thresholds for conflict stemming from autonomous weapons will "shift the burden of armed conflict from soldiers to civilians in battle zones."[62] Wars inevitably produce some harm to civilians, and thus if use of autonomous systems increases military actions, tolls on civilian populations can potentially be greater. More use of military options also creates risk that conflicts will escalate.

Escalation Management

Per discussion of the risk of accidents of AI systems and unpredictable emergent effects, some have noted that military AI creates the risk of a "Flash War" that neither party to the conflict intended. As autonomous systems are more regularly deployed, perhaps in closer proximity to adversaries employing their own autonomous systems, there is a risk that military actions will be executed not just rapidly but at machine speed. The space for deliberate diplomatic negotiations will potentially decrease, increasing the risk of miscalculation and misunderstanding, and thus leading to the possibility of rapid, inadvertent, and accidental escalation. Due to the emergent interaction effects of AI systems, it will be hard to predict how conflicts might intensify, and escalation will be harder to manage.

Proliferation

Much of the R&D of AI is conducted by the private sector and by nongovernment researchers. These efforts have significantly advanced this technology, identified new types of applications, and decreased the costs. Much foundational research is publicly available, and researchers have also created open-source tools to promote wide use.[63] Although such research and tools are not focused on military applications, many of these capabilities are dual- or multiuse across civilian and military contexts. For example, in a futuristic depiction of a malicious use of open-source tools, the "Slaughterbots" video depicts the combination of three technologies: small UAVs, facial recognition technology, and a munition. Less mature versions of all of these technologies are already commercially available. As these types of capabilities improve and costs decrease, they will be more readily accessible for actors outside of traditional global powers. The 2015 "Open Letter from AI & Robotics Researchers" notes that "autonomous weapons will become the Kalashnikovs of tomorrow" and that "it will only be a matter of time until they appear on the black market and in the hands of terrorists."[64]

---

[62] HRW and International Human Rights Clinic, 2012, p. 39.

[63] For instance, Google's TensorFlow is an open-source toolset and library. See "An Open Source Machine Learning Framework for Everyone," TensorFlow, n.d.

[64] Future of Life, 2015.

This is a dire warning of proliferation from some of the experts closest to the technology.

The spread of cyber capabilities demonstrates these risks. It is not only the traditional military powers that have sophisticated cyber tools. Now more than 30 countries have military cyber programs, and North Korea and Iran have already leveraged their cyber programs for a variety of malicious purposes.[65] In addition, criminals have also taken advantage of readily available cyber tools to steal identities, money, and information. This proliferation is especially troubling, since these systems enable small-scale malicious actors to inflict significant damage on even well-defended targets.

Strategic Stability

A final strategic risk is that as AI-enabled tools develop, the basic principles that have ensured relative stability among the global powers since World War II are weakened. In particular, AI-enabled systems might become advanced to the point that they undermine "second-strike" capabilities that are essential to deterrence of nuclear war through the principle of mutual assured destruction. A recent RAND report has considered the possibility that, as AI improves, it might be used to locate all of an adversary's nuclear launchers.[66] With this capability, an aggressor could attack without fear of nuclear retaliation. Even the perception that nuclear launchers would be vulnerable in this way might encourage a state to undertake a first strike to preempt the possibility that they might lose the ability to use nuclear weapons later in a conflict. Such a scenario could be highly destabilizing and put the entire world at risk of nuclear catastrophe.[67]

Perspectives on Mitigating Risks of Military Artificial Intelligence

An overarching concern among technologists, advocates, and other parties is that military establishments will hasten to integrate AI without paying sufficient regard to the seriousness of these risks. As states compete to attain the greatest the military benefits of AI, they might not put proper precautions in place. This dynamic might be characterized as a "race to the bottom" where military applications of AI will result in less overall security.

---

[65] See, for instance, David Sanger, *The Perfect Weapon: War, Sabotage, and Fear in the Cyber Age*, London: Scribe Publications, 2018; David E. Sanger, "U.S. Indicts 7 Iranians in Cyberattacks on Banks and a Dam," *New York Times*, March 24, 2016; Antonio DeSimone and Nicholas Horton, *Sony's Nightmare Before Christmas: The 2014 North Korean Cyber Attack on Sony and Lessons for US Government Actions in Cyberspace*, National Security Report, Laurel, Md.: Johns Hopkins Applied Physics Laboratory, 2017.

[66] Edward Geist and Andrew J. Lohn, *How Might Artificial Intelligence Affect the Risk of Nuclear War?* Santa Monica, Calif.: RAND Corporation, PE-296-RC, 2018.

[67] For additional discussion of AI risks to strategic stability, see Michael C. Horowitz, "Artificial Intelligence, International Competition, and the Balance of Power," *Texas National Security Review*, Vol. 1, No. 3, May 2018, pp. 36–57; Vincent Boulanin, "The Promise and Perils of Artificial Intelligence for Nuclear Stability," *Our World*, December 7, 2018.

Some stakeholders have articulated proposals to mitigate the risks of military AI, and we provide an overview of some here. The greatest concern about military AI is focused on weapon systems that do not have a robust role for a human operator to authorize, control, monitor, intervene, or otherwise have significant involvement with the system. For instance, proponents of a new LAWS treaty focus on prohibiting systems that select and engage targets without human judgment and control. Advocates and governments have argued that the key to minimizing most risks is maintaining some level of human agency over these systems.[68] Thus, the major issues regarding mitigating AI risks revolve around three questions: What role must humans play in military AI systems? How should military AI systems be designed, tested, and managed to mitigate risks? And what can governments do, individually or cooperatively, to ensure that humans maintain sufficient agency?

Human Role in Military Artificial Intelligence Systems

Human in the Loop

As explained in Chapter 2, the role of human operators in AI applications is often discussed in terms of their relationship with the canonical "OODA loop." Many have emphasized the importance of a human in the loop to mitigate ethical and operational risks of military AI. With a human in the AI system's loop, that person can ensure that the system complies with applicable laws and rules of engagement and can be held accountable for the system's actions if it does not. The human can also be the moral focal point responsible for protecting human dignity, as well as the source of emotions that guide just humanitarian conduct. And although humans make mistakes, they generally do not make the same types of mistakes as autonomous systems. Similarly, although humans can be compromised, they are generally not compromised in the same way or through the same methods as a computer system. Thus, humans can also provide an additional check to support the proper deployment and use of an autonomous system.

However, in some military applications, there is an incentive to move humans to positions "on" the loop. For instance, in the context of defensive systems that need to react quickly to incoming threats, human authorization for every engagement might slow the system down and undercut threat-defeating capabilities. Especially in contexts where an adversary is leveraging its own high-speed AI systems, human-in-the-loop approaches risk a competitive disadvantage. In these cases, there is pressure to move the human operator to a position of monitoring the systems

---

[68] With respect to governments, see "Emerging Commonalities, Conclusions and Recommendations," Group of Governmental Experts of the High Contracting Parties to the Convention on Prohibitions or Restrictions on the Use of Certain Conventional Weapons Which May Be Deemed to Be Excessively Injurious or to Have Indiscriminate Effects, Geneva, August 2018, p. 3, which states: "Human responsibility for the use of force must be retained." Also see United Nations Office at Geneva, "Chair's Summary of the Discussion," Group of Governmental Experts of the High Contracting Parties to the Convention on Prohibitions or Restrictions on the Use of Certain Conventional Weapons Which May Be Deemed to Be Excessively Injurious or to Have Indiscriminate Effects, Geneva, April 2018, p. 5, which states: "Delegations reaffirmed the essential importance of human, control, supervision, oversight or judgment in the use of lethal force."

and ensuring a mechanism where intervention is a possibility. However, AI systems with human-on-the-loop modes raise questions about the extent to which a human can actually intervene rapidly enough to curtail an engagement. There will also be contexts in congested electromagnetic environments in which communications degradation or denial means that the human might lose the possibility to effectively monitor the system's operation and intervene.

The loop concept provides an approach to mitigating risk, but it is incomplete. First, the loop applies most readily to autonomous weapon systems involved in selecting and engaging targets. However, as discussed in Chapter 2 there are many other types of AI military applications, such as AI decision support systems or logistics systems, and it is unclear in those cases what role the human in the loop plays to mitigate risks. In addition, the loop concept is not sufficient to ensure human involvement in other parts of the system life cycle, such as robust testing and evaluation, weapons review, and other steps where risk mitigation is important. Further, there are interpretive challenges regarding the size of the relevant loop. For systems that can maintain presence over time, or "loiter," and conduct multiple strikes, a human authorizing initial deployment of the system might be considered to be "in the loop," even as the system undertakes a range of actions over prolonged periods. Last, there are different types of C2 arrangements and mechanisms for intervention, and the loop concept does not itself help distinguish these different varieties. Consequently, although the loop concept is useful, there are other risk-mitigation approaches that may be important.

Meaningful Human Control and Appropriate Human Judgment

Another concept that seeks to mitigate risks by specifying the role of human involvement in an autonomous system is that of *meaningful human control*. The idea here is that humans must play a significant moral, legal, and operational role over the AI systems by administering control. Heather Roff and Richard Moyes have argued that meaningful human control not only is relevant on deployment of the system but also needs to be embedded throughout phases of a system's development and deployment life cycle, starting with design, development, and training in an ante bellum period and continuing human control during attacks and structures of accountability post bellum.[69]

DoD has emphasized a different concept, that of *appropriate human judgment*. Directive 3000.09, states that "autonomous and semi-autonomous weapon systems shall be designed to allow commanders and operators to exercise *appropriate levels of human judgment* over the use

---

[69] Heather Roff and Richard Moyes, "Meaningful Human Control, Artificial Intelligence, and Autonomous Weapons," Briefing Paper for Delegates at the Convention on Certain Conventional Weapons (CCW) Meeting of Experts on Lethal Autonomous Weapons Systems (LAWS), Geneva, April 2016. Other commentators have also written on this concept; see Michael Horowitz and Paul Scharre, "Meaningful Human Control in Weapon Systems: A Primer," CNAS Working Paper, Center for a New American Security, March 2015.

of force."[70] In a statement to the 2016 UN CCW Informal Meeting of Experts, the U.S. delegation noted that the United States prefers this formulation over meaningful human control since the latter is "subjective and thus difficult to understand."[71] The U.S. statement argued that the concept of appropriate human judgment better captures the importance of the human-machine relationship "throughout the development and employment of a system and is not limited to a moment of a decision to engage a target."[72] However, the concept is also somewhat nebulous and more is needed to define "appropriate."

In the August 2018 UNCCW GGE, the U.S. delegation went further in criticizing the concept of meaningful human control by arguing that it "risks obscuring the genuine challenges in human-machine interaction"[73] The delegation also clarified that the ultimate goal of appropriate human judgment is to "help effectuate the intent of commanders and the operators of weapons systems."[74] Rather than decrease human control, the U.S. delegation argued that increased autonomy in weapon systems can actually better effectuate commanders' intentions, and thus a ban on all LAWS would risk undercutting humanitarian values that can be promoted by these technologies.[75]

Other commentators have come out with their own terminology to best describe the type of human involvement required in the use of autonomous weapons—a debate that has become known as the "X human Y" debate (where X could be "meaningful," "appropriate," "sufficient," or some other term, while Y could be "control," "judgment," "agency," and so forth). Whatever terms one ultimately supports, Roff helpfully notes that "there is consensus that no one wants weapons that operate *out of human control*."[76]

System Life Cycle Approach

An additional detail regarding human involvement concerns the phases of the systems life cycle in which humans must play a role. In the most recent August UN CCW GGE meeting, the

---

[70] Executive Services Directorate (ESD), DoD Directive 3000.09, *Autonomy in Weapon Systems*, November 21, 2012, revised May 8, 2017, p. 2 (emphasis added).

[71] U.S. Mission, Geneva, "U.S. Delegation Opening Statement (as delivered)," The Convention on Certain Conventional Weapons (CCW) Informal Meeting of Experts on Lethal Autonomous Weapons Systems (LAWS), April 2016.

[72] U.S. Mission, Geneva, 2016.

[73] U.S. Mission, Geneva, "U.S. Delegation Statement on Possible Options," The Meeting of the Group of Governmental Experts of the High Contracting Parties to the CCW on Lethal Autonomous Weapons Systems, August 2018b.

[74] U.S. Mission, Geneva, "U.S. Delegation Statement on Human-Machine Interaction," The Meeting of the Group of Governmental Experts of the High Contracting Parties to the CCW on Lethal Autonomous Weapons Systems, August 2018a.

[75] This is a U.S. government position. We do not definitively evaluate this claim.

[76] Roff, 2016, p. 2 (emphasis in original).

United Kingdom distinguished different stages of the life cycle and made a case for human touchpoints and control at each stage.[77] Figure 3.2 illustrates this concept.

Figure 3.2. United Kingdom Framework for Considering Human Control Throughout the Life Cycle of a Weapon System

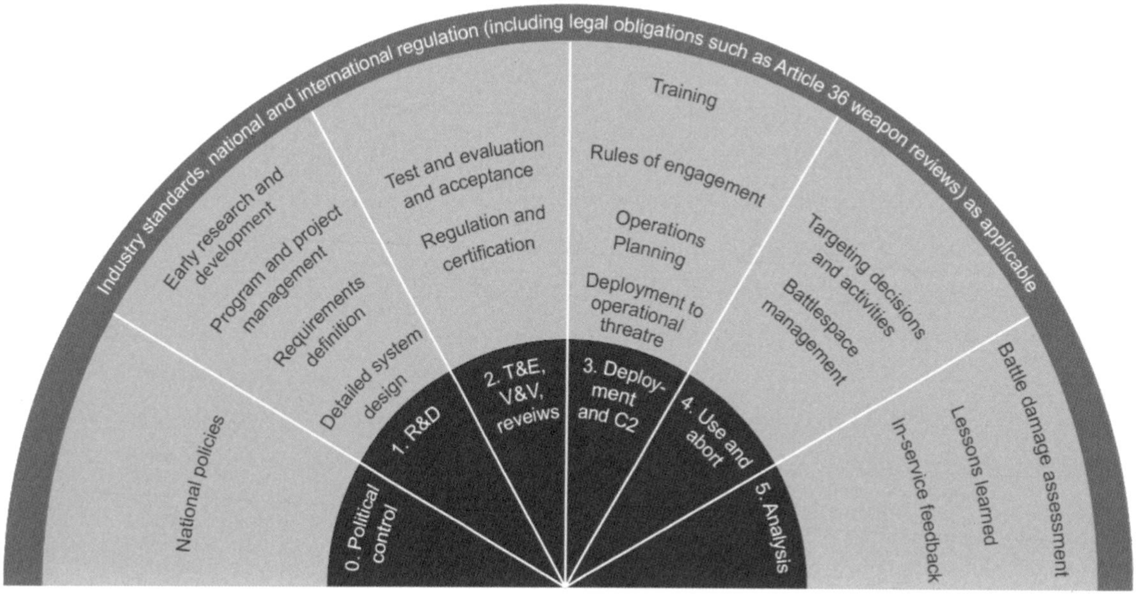

NOTE: T&E = testing and evaluation of the system; V&V = validation and verification that the system operates as intended.

As the figure indicates, this approach emphasizes that human-led risk mitigation starts even before a system is developed, through the government's national policies (for instance, DoD Directive 3000.09), and extends throughout the life cycle, all the way to battle-damage assessment and other postemployment analyses. This approach illustrates that human involvement is a multifaceted phenomenon, with humans playing distinct roles at different times, and not just in the target-selection and engagement process. If governments find this risk-mitigation approach valuable, they will need to carefully consider the variety of human touchpoints throughout the system's life cycle, including within elements of system design and development, rules of engagement, and the targeting process.

---

[77] "Human-Machine Touchpoints: The United Kingdom's Perspective on Human Control over Weapon Development and Targeting Cycles," Paper Submitted by the United Kingdom to Group of Governmental Experts of the High Contracting Parties to the Convention on Prohibitions or Restrictions on the Use of Certain Conventional Weapons Which May Be Deemed to Be Excessively Injurious or to Have Indiscriminate Effects, August 2018.

System Design and Development

In this section, we discuss several key criteria relevant to ensuring human involvement in military AI to mitigate risks. These criteria are not meant to be comprehensive, but they do indicate the type of relevant system design issues that need to be addressed.

Constraints on Time Frame

Constraints should be set on the length of time autonomous systems can be deployed without human involvement or direction. As discussed above, certain autonomous systems are intended to loiter or remain active for prolonged periods. However, DoD Directive 3000.09 emphasizes that autonomous weapons must "complete engagements in a timeframe consistent with commander and operator intentions, and if unable to do so, terminate engagements or seek additional human operator input before continuing the engagement."[78] This requires that a constraint be set on the length of time a system could be active. However, the details of this constraint will depend on system features, commander's intention, the operational environment, and other contextual factors.

Constraints on Geography

Some autonomous systems might have extensive navigational capabilities to move in space. A geographical constraint limits the extent of geographical movement that the system can undertake autonomously. To mitigate risk associated with unconstrained movement, operators must be able to set parameters within which the system must remain. If events or malfunctions result in the system exceeding its geographical constraints, it might be programmed to terminate the mission and return to base.

Constraints on Types of Tasks

Although most current autonomous systems tend to have only a narrow band of functionality, future technological developments might produce systems that can take on a wide variety of tasks. This constraint would limit the number and types of tasks the system can perform. Systems that take lethal action or certain other consequential risks, might involve additional controls than more benign systems. For instance, if the system has the capability to target persons and not just military objects, this would seem to require greater operational constraints and testing.

Reliability and Predictability

Standards of reliability and predictability need to be established to ensure the system will operate according to the intention of the design. Such standards need to be set in the design and development stage of system production, and systems should be required to meet them before being declared operational. Once systems are operational, data need to be continuously collected

---

[78] ESD, 2017, p. 2.

and periodically evaluated to ensure they continue to meet standards. Again, more reliability and predictability would presumably be required for more consequential systems.

Accessible Information

An additional criterion concerns the degree of system transparency that is required. In order to mitigate some AI risks, the operator must have knowledge of what the system is intended to do in a specific operational context and must be able to determine within a reasonable time frame how the system arrived at critical decisions or why it took particular actions. According to a white paper the United States delegation submitted to the 2017 UN CCW GGE, a best practice for the interface between people and machines for autonomous and semiautonomous weapons is to "(1) be readily understandable to trained operators; (2) provide traceable feedback on system status; and (3) provide clear procedures for trained operators to activate and deactivate system functions."[79] Similarly, the chair's summary of the April 2018 UN CCW GGE meeting says the operator should "know the characteristics of the weapons system, [be] assured that [they] are appropriate to the environment in which it would be deployed and [have] sufficient and reliable information on them in order to make conscious decisions and ensure legal compliance."[80] However, as discussed above in the section on operational risk related to trust and reliability, designing and building AI systems with sufficient transparency is a serious challenge.

Options for Intervention

Given the risks associated with accidents, emergent effects, or other problems, an important system design criterion concerns the possibility of intervention to redirect or stop the system. To the extent possible with the parameters of mission requirements, the human operator should be able to intervene in the system's actions in a timely manner to redirect it as necessary. How timely such options for intervention must be will depend on context. Operators should be able to intervene in the actions of systems employing lethal force in offensive operations in proximity to noncombatants the most quickly; conversely, time constraints for system intervention may be more relaxed for systems not employing lethal force and those taking defensive actions in areas far removed from noncombatants.

International Regulatory Options

This section describes some of the options for international action to mitigate risks.

---

[79] "Autonomy in Weapon Systems," U.S. submission to the Meeting of the Group of Governmental Experts of the High Contracting Parties to the Convention on Prohibitions or Restrictions on the Use of Certain Conventional Weapons Which May Be Deemed to Be Excessively Injurious or to Have Indiscriminate Effects, Geneva, November 10, 2017, p. 3.

[80] UNOG, 2018c.

New Treaty

Advocates have called for a new law that would prohibit autonomous weapons that select and engage targets without human involvement. Some states have supported this call; however, these are predominantly states that do not have advanced military capabilities (China being the one exception). Others have argued that existing international law, especially LOAC, is sufficient to protect against the potential humanitarian consequences of military AI. Despite ongoing UN discussions, a ban or other regulation on AI in military applications is not likely in the near term. The states developing these systems see great military and humanitarian utility in them, and there is a general desire not to slow technological development or otherwise interfere with the integration of AI for military use.

Moreover, in all arms control agreements there are concerns about verifiability and reciprocity. These concerns are greater in the field of military AI than in most other categories of weapons, because being largely resident in computer software, AI is not transparent to treaty monitors or other states. As explained in Chapter 2, the difference between a semiautonomous and fully autonomous system is often just a software setting. So even if a new ban is established, there will be serious concerns about compliance.

Confidence-Building Measures

In the absence of a new treaty, states may be able to undertake confidence-building measures to promote transparency and risk-reduction. Building confidence and transparency in this area will be challenging, because much R&D is classified, but embracing best practices in weapons review might be a productive place to start. For instance, Article 36, a UK-based organization, provides guidance for undertaking national legal reviews, which involve "evaluating systems to ensure meaningful human control." Such reviews would seek to ensure systems are predictable, reliable, and have explainable technology. They would look for the ability to provide accurate information for use and context of use, the capability for timely human intervention, and some standard of accountability.[81] Encouraging international endorsement of such standards could be a first step toward developing a climate in which a formal agreement incorporating them is reached.

Another approach to developing confidence-building measures might include seeking international agreement on standards for testing, evaluation, and explicability. In this area, the U.S. white paper submission to the CCW maintains that "rigorous and realistic testing standards and procedures can ensure that commanders and national security policymakers can have a reasonable expectation of the likely effects of employing the weapon in different operational contexts."[82] It may be possible to develop international best practices around testing and safety

---

[81] Article 36, "Autonomous Weapon Systems: Evaluating the Capacity for Meaningful Human Control in Weapon Review Processes," Discussion paper for the Convention on Certain Conventional Weapons (CCW) Group of Governmental Experts meeting on Lethal Autonomous Weapons Systems (LAWS), November 2017, p. 4.

[82] UNOG, 2017.

standards, since these help to mitigate risk in areas that are shared by all nations developing these technologies.

Political Commitment

Finally, seeking a political commitment from states developing military applications of AI might be a positive step in efforts to regulate these technologies. A political commitment would be a declaration by national leaders affirming that they commit to developing and using military AI subject to specified criteria. This would not be a legally binding treaty but a political declaration, similar to commitments by countries to abide by certain norms of responsible behavior in cyberspace. The chair's report from the 2018 UN CCW GGE meeting notes that a political declaration might affirm "that humans should be responsible for (a) making final decisions with regard to lethal force and (b) maintaining control over autonomous weapons systems, without prejudice to policy outcomes."[83]

Being easily made and easily abandoned, such declarations carry less weight than formal agreements. Nevertheless, they are positive steps that could encourage states to engage in other confidence-building measures.

Conclusion

AI has the capacity to fundamentally alter the character of war. Although defense officials in the United States and other countries see significant benefit in AI, technologists, advocates, and state governments have been outspoken in raising concerns. This chapter has described the ethical, operational, and strategic risks these groups have raised. A worrisome overarching fear is that as states feel increasing pressure to integrate AI into military applications, there will be a "race to the bottom" where these risks will not be properly addressed. This line of thought holds that states will hasten to develop and use novel technologies without putting the necessary humanitarian and safety constraints in place. To address some of the risks of military AI, there is a developing consensus that humans need to maintain control over the development, deployment, and use of military AI systems. However, there are still many open questions about the degree of control needed, what form it should take, and how such safeguards can be enforced domestically and internationally. This chapter has described some of the options for mitigating risks, but more research is needed to explore how key actors can find areas of shared interest to ensure that the risks of military AI are adequately addressed.

---

[83] UNOG, 2018c, p. 12.

4. Military Artificial Intelligence in the United States

Despite a history of developing offensive military applications of AI, the official U.S. policy position on these emerging technologies has been one of restraint. This can be seen in many statements from U.S. leaders, such as this one from then–Deputy Secretary of Defense Robert Work: "Autonomy will be used only to empower humans, not to make individual or independent decisions on the use of lethal force."[1]

This chapter will summarize the history of military AI development in the United States, describe some capabilities recently developed and projected to be available in the near future, and review policies and risk-mitigation approaches that exist, or are possible from the U.S. perspective, with respect to military uses of AI.

Brief History of Military Artificial Intelligence Development in the United States

The component technologies to provide various aspects of autonomy have been available for several decades. The United States has a long history of attempting to develop these capabilities and has even fielded a number of autonomous or semiautonomous weapon systems. This section offers a brief summary of the defensive and offensive systems developed to date, as well as systems already available to support planning and logistics functions.

Defensive Systems

One of the most often discussed weapon systems with autonomous capabilities is the Aegis Ballistic Missile Defense System, which was introduced in 1983. Ships equipped with Aegis Ballistic Missile Defenses are capable of performing a number of functions autonomously.[2] Operators can configure them to respond to many different threats by selecting from, or even combining, a set of specified "doctrines" labeled "Semi-Auto," "Auto SM," and "Auto-Special."[3] Both Semi-Auto and Auto SM are semiautonomous—they are human-in-the-loop configurations. Auto-Special, however, is supervised autonomous, in which the human is only on the loop. This doctrine is intended for cases in which threats exceed the operator's ability to coordinate and manage defenses. Switching between doctrines can be done with the push of a button, demonstrating the ease with which it is possible to convert between semiautonomous and supervised autonomous systems. Similar in role to the Aegis Combat System is the Phalanx

---

[1] DoD, "Remarks by Deputy Secretary Work on Third Offset Strategy," Brussels, April 28, 2016.

[2] DOT&E, "Aegis Ballistic Missile Defense System," *FY17 Ballistic Missile Defense Systems*, 2017, pp. 291–296.

[3] Scharre, 2018, pp. 163–169.

Close-In Weapons System (CIWS), which was approved for production in 1978. It can also be operated in autonomous mode to defend against salvos of missiles or high numbers of attacking aircraft.[4]

Offensive Systems

It can be argued that defensive systems, such as those described above, present the most compelling case for autonomy, because of their potential to face an onslaught of attacks that exceed the operators' ability to manage. But U.S. developers have also produced autonomous weapons that are more offensively focused. The Tomahawk Anti-Ship Missile (TASM) is an early example of a fully autonomous weapon system. It was designed to loiter and search for ships to target then engage those targets once detected.[5] Configured to operate in that mode for a short period in the early 1990s, the system was never used in that way and was taken out of service due to concern that it would lead to accidents by engaging unintended targets. The program has since been resumed to compete with the more contemporary Long Range Anti-Ship Missile (LRASM).[6]

Around the same time period another semiautonomous weapon, the High-Speed Anti-Radiation Missile (HARM), was introduced to attack radiation signals such as those emanated by air defense systems. These missiles are semiautonomous because, although they seek their own targets, those targets generally have to have been previously identified by an operator. However, some variants from other countries have longer loiter times and are therefore described as fully autonomous systems. Unlike TASMs, HARMs have been popular weapons and have been used extensively in combat.

More recent examples of advances in AI and autonomy include the products of DARPA's Joint Unmanned Combat Air Systems program. These autonomous flying systems have waxed and waned in popularity and programmatic support, but DARPA has managed to produce an impressive demonstration vehicle in the X47-B. In 2013, the X47-B successfully completed an autonomous landing on an aircraft carrier.[7] Following that demonstration, in 2015, the X47-B successfully conducted an autonomous aerial refueling.[8] Despite the success of these demonstrations, the X47-B program was canceled due to Navy concerns about cost and lack of

---

[4] Kyle Mizokami, "Phalanx: The US Navy's Last-Ditch Automated Air Defense System," *Popular Mechanics*, April 14, 2016.

[5] Scharre, 2018, p. 49.

[6] Sydney J. Freedberg Jr., "Tomahawk vs LRASM: Raytheon Gets $119M for Anti-Ship Missile," *Breaking Defense*, September 11, 2017.

[7] Nidhi Subbaraman, "After Two Historic Carrier Landings, Navy's X47-B Drone Scrubs a Third," *NBC News*, July 11, 2013.

[8] Kris Osborn, "Navy Conducts First Aerial Refueling of X47-B Carrier Launched Drone," Military.com, April 22, 2015.

stealthiness, but the technologies it exhibited in unmanned autonomous aerial refueling are likely to continue with the development of the MQ-25 and future systems.[9]

Artificial Intelligence for Planning and Logistics

Most of the tasks that military forces conduct are related to aspects of war other than conducting or defending against kinetic strikes. There is a wide variety of other categories of tasks, such as personnel management, intelligence, logistics, communications, and planning, in which AI can be used and has been developed and fielded.

For instance, the Dynamic Analysis and Replanning Tool (DART) assisted in developing plans for moving troops and equipment from Europe to Saudi Arabia during Operation Desert Shield and Operation Desert Storm from 1990 to 1991.[10] The techniques that made that tool possible are unlikely to be considered AI by today's standards, but at the time, they were, and the project was deemed an overwhelming success. Victor Ries, then-director of DARPA, concluded that DART had single-handedly paid back the 30 years of DARPA investment in AI at the time.[11]

About a decade later, a much more sophisticated tool for planning force deployment and solving logistics problems was developed. The Joint Assistant for Deployment and Execution (JADE), a project supported by the Air Force Research Laboratory (AFRL) and DARPA, was designed to build a preliminary force deployment plan, including the Time-Phased Force Deployment Data.[12] It was first demonstrated in 1999 for U.S. Southern Command, which was responsible for responding to Hurricane Mitch.[13]

DARPA also explored the possibility of using AI to assist in even more critical planning with the Survivable Adaptive Planning Experiment (SAPE), which was tested in part in 1991.[14] Whereas JADE was designed for planning force flows for conventional conflicts, SAPE was intended for generating nuclear war plans. It was supposed to be able to rewrite the Single Integrated Operational Plan in three days, a task that normally took 18 months.[15] And it was

---

[9] Tyler Rogoway, "We Finally See the Wings on Boeing's MQ-25 Drone as Details About Its Genesis Emerge," *The Drive*, March 13, 2018.

[10] Sara Reese Hedberg, "DART: Revolutionizing Logistics Planning," *IEEE Intelligent Systems*, Vol. 17, 2002, pp. 81–83. DOI: 10.1109/MIS.2002.1005635

[11] Hedberg, 2002.

[12] Alice M. Mulvehill and Joseph A Caroli, "JADE: A Tool for Rapid Crisis Action Planning," paper presented at the Command and Control Research and Technology Symposium, United States Naval War College, Newport, RI, 1999.

[13] Alice M. Mulvehill, Clinton Hyde, and Dave Ranger, "Joint Assistant for Deployment and Execution (JADE)," Air Force Research Laboratory Technical Report: AFRL-IF-RS-TR-2001-171, August 2001.

[14] Edward M. Geist, "It's Already Too Late to Stop the AI Arms Race—We Must Manage It Instead," *Bulletin of the Atomic Scientists*, Vol. 72, No. 5, 2016, pp. 318–321.

[15] The Single Integrated Operational Plan was the Strategic Air Command's nuclear weapons targeting plan.

intended to reduce the time for retargeting strategic weapons from eight hours to three minutes, which would make it possible to perform retargeting during an attack.[16] This project ended when the demise of the Soviet Union eliminated the principal adversary that would have been the target of SAPE's planning. Errors in these nuclear planning systems could have had disastrous consequences. Although this program was canceled, there are many other parts of the nuclear enterprise where incorporation of AI can be envisioned, the risks of which are just starting to be explored.[17]

For instance, the field of strategic missile warning and attack assessment has employed concepts of AI in the past and will do so increasingly in the future. Detection of missile launches and assessment of the nature and magnitude of the threats they present to the homeland are central to the national strategic warning mission. False alarms in this critical function and near disasters occurred on both sides during the Cold War.[18] An erroneous assessment of nuclear attack might be the most dramatic example of the risk of failures from unexpected behavior of AI-enabled systems.

Mistakes and Near Misses Lead to Caution

Despite the long history of military investment in AI and autonomy, the United States does not have many autonomous, or even semiautonomous, systems in use. U.S. leaders and military officers at all levels are skeptical about the reliability and utility of these systems. This has often caused even promising programs, such as most of those described above, to be canceled, dropped upon completion, or unused in deployment. One aspect driving this reticence is likely a by-product of the long history the United States has with this technology and the many accidents or near misses that are imprinted on the military's collective memory.

Most of these incidents were later determined to have been caused by human error, even when involving weapons with varying degrees of autonomous capability. For example, the downing of Iran Air Flight 655 by the USS *Vincennes* in 1988 in the Strait of Hormuz, was by missiles launched from the Aegis system, but they were fired by a human operator. Still, incidents such as that one remind commanders and policymakers of the risks inherent in these systems, and there are plenty of cases in which the incidents were less clearly caused by human error, because the systems involved were operating in more autonomous modes. Such incidents tend to entrench fears of autonomous systems in the minds of U.S. leaders.

---

[16] Alex Roland and Philip Shiman, *Strategic Computing: DARPA and the Quest for Machine Intelligence, 1983–1993*, Cambridge, Mass.: MIT Press, 2002, p. 305.

[17] Geist and Lohn, 2018.

[18] In one of the most noted examples, on September 26, 1983, the Soviet satellite warning system generated an attack alarm reporting five missiles heading from the United States to the Soviet Union. Fortunately, Soviet Lieutenant Colonel Stanislav Petrov, the missile warning officer on duty, chose to delete the reports rather than send them to higher echelons of command. See Marc Bennetts, "Soviet Officer Who Averted Cold War Nuclear Disaster Dies Aged 77," *The Guardian*, September 18, 2017.

During the Gulf War in 1991, the USS *Missouri* was believed to be under attack from an Iraqi Silkworm Missile and fired its chaff in defense. At the same time, a Phalanx CIWS system aboard the nearby USS *Jarrett*, operating in an autonomous target-acquisition mode, detected the chaff and fired at it.[19] Four rounds from the USS *Jarrett* hit the USS *Missouri*. Fortunately, this incident did not result in injuries.

Another notable incident involving a Phalanx system occurred in 1996 during a multinational training exercise in the Pacific. As part of the exercise, a Navy A-6E Intruder was towing a target plane that was to be shot down by Japanese participants. Instead of locking onto the target plane, however, the Phalanx locked onto the Intruder and opened fire.[20] The pilots ejected and survived, but the plane was destroyed.

Patriot missiles have been involved several similar incidents. The incidents are described as the results of automation bias, in which the humans on the loop trusted the supervised autonomous systems too much, allowing them to make catastrophic mistakes. In one example, in 2003, a U.S. Patriot battery shot down a Tornado flown by the Royal Air Force, killing two of the crew.[21] In another Patriot incident, also in 2003, a U.S. Navy aircraft was mistaken for an Iraqi missile, and the pilot was killed.[22]

In addition to these military incidents, there has been a growing number of deadly accidents on public streets involving the use of autonomous vehicles. In important ways, autonomous vehicles face a simpler environment than autonomous military systems do. For one thing, the goals of all parties in the environment in which autonomous vehicles operate are aligned—they all want to transport people and items safely. For another, the environment has been carefully designed with common rules that are intended to increase the safety and ease of transportation. Neither of those conditions exist for AI in weapon systems, as commanders and military acquisition officials are keenly aware. It is with that perspective in mind that they view the accidents and fatalities in autonomous vehicles. This contributes to their trepidation about the risks of fielding military autonomy.

Summary of Current Capabilities and Future Projections

While commanders and leaders in the acquisition community have been hesitant to deploy autonomous weapon systems, such reservations have never been as strong among the individuals in the research community who develop these technologies. As a result, research organizations are developing a wide range of impressive applications of AI and autonomy. In fact, interest in

---

[19] David Hambling, "Iran's 'New' Anti-Missile Artillery," *WIRED*, May 27, 2009.

[20] Kevin Sullivan, "Japanese Ship Downs U.S. Plane," *Washington Post*, June 5, 1996.

[21] David Axe, "That Time an Air Force F-16 and an Army Missile Battery Fought Each Other," *Medium*, July 5, 2015.

[22] Thomas E. Ricks, "Investigation Finds US Missiles Downed Navy Jet," *Washington Post*, December 11, 2004.

this area appears to be growing. One of the clearest signs that this is true has been the recent establishment of the Defense Innovation Unit in the heart of Silicon Valley, an organization with the mission of "contracting with companies offering solutions in a variety of areas—from autonomy and AI to human systems, IT, and space."[23] It and other organizations are aggressively advocating the development of AI and autonomy within DoD. Consequently, a number of new capabilities may be available in the near future. We describe a few of them below.

Long Range Anti-Ship Missiles

We have briefly touched on a few of these applications already in our discussion of systems previously developed. One of them is the LRASM, now being developed and advertised to be able to autonomously select and engage targets, even in GPS- and communications-denied environments.[24] LRASMs will need onboard abilities to manage these engagements in situations where human operators are not able to provide guidance or intervene. Depending on the degree of specificity in instructions that operators can provide the weapon prior to its taking over, the LRASM will operate in at least a semiautonomous mode and may have to be fully autonomous.

Autonomous Flying Vehicles

We have already mentioned progress in the development of unmanned aerial vehicles, such as the X47-B and the MQ-25. AFRL has also been in the process of developing autonomous flying vehicles under the auspices of its Loyal Wingman program.[25] Initiatives in that effort have included outfitting the F-16, which is normally human operated, with autonomous capability. The autonomous F-16 was demonstrated in 2017 when it was able to react dynamically to a changing threat environment and conduct an air-to-ground strike mission.[26] The vision for these systems is for them to operate as extensions of a pilot in another aircraft who can command them to perform dangerous tasks. This is seen as a stepping stone toward developing swarms of autonomous systems that do not require pilots. This concept has some high-level support. For instance, in 2015 Secretary of the Navy Ray Mabus said, "The F-35 should be, and almost certainly will be, the last manned strike fighter aircraft the Department of the Navy will ever buy or fly."[27]

---

[23] Billy Mitchell, "'No Longer Experimental'—DIUx Becomes DIU, Permanent Pentagon Unit," *FEDSCOOP*, August 9, 2018.

[24] Lockheed Martin, "Long-Range Anti-Ship Missile (LRASM)," YouTube, May 2, 2016.

[25] Daniel Wassmuth and David Blair, "Loyal Wingman, Flocking, and Swarming: New Models of Distributed Airpower," *War on the Rocks*, February 21, 2018.

[26] Loren Blinde, "US Air Force, Lockheed Martin, Demonstrate Manned/Unmanned Teaming," *Intelligence Community News*, April 11, 2017.

[27] Sam LaGrone, "Mabus: F-35 Will Be 'Last Manned Strike Fighter' the Navy, Marines 'Will Ever Buy or Fly,'" *USNI News*, April 15, 2015.

As Mabus's comment indicates, the Air Force is not the only service interested in making its systems more autonomous. The Army has demonstrated autonomous helicopters, which it is developing primarily for resupply missions.[28] And though the Navy ultimately canceled its X47-B UCAS program, it has continued pursuing autonomous capabilities in developing a variety of autonomous surface and undersea vessels.[29]

Concepts of operations for collaborative autonomous flying capabilities are far from fully developed, which is also why there is only mixed support for procuring and fielding these systems. But DARPA is pressing ahead with several other programs that, if successful, will provide some of the capabilities needed to make autonomous combat aircraft viable. These programs include Target Recognition and Adaptation in Contested Environments and Collaborative Operations in Denied Environments, and they are intended to demonstrate that autonomous aerial vehicles can work as a team and identify targets in unpredictable environments without the ability to communicate with human operators. These goals are ambitious, though increasingly feasible. But it remains to be seen just how valuable the acquisition community and commanders in the field will find them once they are demonstrated.

Artificial Intelligence for Logistics and Planning

In terms of fielding operational systems in the near term, there has been more interest in using AI to enhance software-based applications than there has in developing autonomous weapon systems. Project Maven, a rapid effort to apply industrial image-recognition and object-detection technology to defense, has received the most attention lately. As mentioned in Chapter 3, the overarching motivation for this program is that the large volume of image and video data being collected in ISR has exceeded the manpower available to view and analyze it. AI is needed to prioritize and filter results as data come in.[30] So far, the results have been impressive. Within just six months from initiation, the project was able to assist in intelligence processing in actual operational use in the fight against the Islamic State of Iraq and the Levant (ISIS).[31] As also mentioned in Chapter 3, while development of Project Maven and similar applications is likely to continue, the road ahead may be rocky, because employees at Google and other companies involved in AI-related projects have applied pressure on management to cease work on such contracts due to their concerns about the militarization of AI.[32]

---

[28] Allen McDuffee, "Black Hawk Drone: Army's Iconic Helicopter Goes Pilotless," *WIRED*, April 30, 2014; Steve Dent, "Autonomous Helicopter Makes First Operational Delivery to Marines," *Engadget*, May 18, 2018.

[29] Kyle Mizokami, "The U.S. Navy Just Got the World's Largest Uncrewed Ship," *Popular Mechanics*, February 5, 2018; Mark Pomerleau, "DoD Plans to Invest $600M in Unmanned Undersea Vehicles," *Defense Systems*, February 4, 2016.

[30] Wakabayashi and Shane, 2018.

[31] Gregory C. Allen, "Project Maven Brings AI to the Fight Against ISIS," *Bulletin of the Atomic Scientists*, December 21, 2017.

[32] Wakabayashi and Shane, 2018.

Yet DoD is likely to persist in its pursuit of AI technologies. To see how strongly the Pentagon believes that the future of warfare involves large-scale data analytics, including the kind described above, one needs only to follow the extent of investment. DoD is currently in the process of selecting among competitors bidding for a contract to supply a large-scale cloud-computing infrastructure. The contract is called Joint Enterprise Defense Infrastructure, and it is estimated to be worth $5 billion a year over two years.[33] In comparison, Amazon's entire cloud computing revenue is about $20 billion annually, and Google's is around $4 billion. While this contract is about far more than AI, the anticipated need for AI is likely fueling demand for such expansive computing infrastructure, which, in turn, will necessitate more AI technologies, such as image processing and natural-language processing, to deal with the volume of data that will become accessible.

Algorithmic Warfare

Algorithmic warfare refers to harnessing the algorithmic systems that underpin AI and ML for employment in future conflicts. This data-centric approach to warfare is in line with the long-term strategy of some of DoD's senior officials. The so-called Third Offset Strategy, which was introduced by Secretary of Defense Chuck Hagel and supported by his successor, Ashton Carter, touted AI and human-machine teaming as a means of maintaining (or, perhaps, regaining) overmatch against near-peer nations.[34] Although the expression "Third Offset Strategy" has since been discarded, the concept behind it is apparent in the formation the Pentagon's Algorithmic Warfare Cross-Functional Team, the organization responsible for Project Maven.[35] DoD's commitment to AI and algorithmic warfare is also apparent in the stand-up of the Joint Artificial Intelligence Center, with the goal of accelerating the delivery of AI-enabled capabilities, scaling the department-wide impact of AI, and synchronizing DoD AI activities to expand Joint Force advantages.[36]

Algorithmic warfare is a broad concept that extends beyond intelligence and data-analysis applications. It can be seen in a number of initial demonstrations by AFRL, including one of an AI system that was able to defeat expert human pilots in simulated dogfighting.[37] Other emerging applications are nonkinetic in nature. One example that is still in the early research stages, but is expected to improve in the coming years, is related to cyber warfare. Some of the

---

[33] Kate Conger, "The Fight for a Massive Pentagon Cloud Contract Is Heating Up," *Gizmodo*, May 8, 2018.

[34] James R. McGrath, "Twenty-First Century Information Warfare and the Third Offset Strategy," *Joint Force Quarterly*, Vol. 82, No. 3, 2016, pp. 16–23.

[35] Adin Dobkin, "DoD Maven AI Project Develops First Algorithms, Starts Testing," *Defense Systems*, November 3, 2017.

[36] Patrick Tucker, "The Pentagon Is Building an AI Product Factory," *Defense One*, April 19, 2018.

[37] M. B. Reilly, "Beyond Video Games: New Artificial Intelligence Beats Tactical Experts in Combat Simulation," *UC Magazine*, June 27, 2016; Chris Baraniuk, "AI Fighter Pilot Wins in Combat Simulation," *BBC News*, June 28, 2016.

experts we interviewed expect that algorithms will be able to discover and exploit vulnerabilities in enemy cyber networks, while detecting and expelling intruders from friendly networks. AI has already been incorporated in commercial antivirus systems.[38] Hoping to further develop these capabilities, DARPA sponsored a Cyber Grand Challenge in 2016. In that event, independent computers sought to hack and to defend themselves autonomously, and they achieved some degree of success: a team from Carnegie Mellon University took home the $2 million prize when their system defeated six other team's systems.[39]

U.S. Policies to Mitigate Risks

As previously mentioned, the United States has shown restraint in the acquisition of autonomous weapons. The reluctance to proceed down that road is reflected in statements by high-ranking military officers, such Vice Chairman of the Joint Chiefs of Staff Paul Selva, who said, "I don't think it's reasonable for us to put robots in charge of whether or not we take a human life."[40] But there are also numerous instances in which AI has been incorporated in actual fielded systems, and there will certainly be more to come. In light of that, it is important to examine the steps that U.S. developers and military professionals have taken to mitigate the serious risks associated with military AI.

U.S. Military Researchers Care About Safety

Among the primary options for risk mitigation is a focus on safety in the design and testing of the systems themselves. Researchers in the United States take this requirement seriously, and there are many programs designed to increase, verify, or validate the safety of autonomous weapons.

DARPA has several programs focused on providing these safety assurances, one of which is aptly named "Assured Autonomy."[41] That program's intent is to develop the technological capability to continually ensure that learning systems maintain safety and functional correctness, even while they are learning and adapting without human supervision or involvement in their training.

As discussed in Chapter 3, among the challenges that learning systems present in terms of safety and verification is that, once they are trained, it can be difficult or impossible for operators to understand how these systems arrived at critical decisions or why they are taking particular actions. To help ensure the safety and functional correctness of AI systems, DARPA has a

---

[38] Alfred Ng, "Microsoft Is Building a Smart Antivirus Using 400 Million PCs," *CNET*, June 27, 2018.

[39] Coldeway, 2016.

[40] Ryane Browne, "US General Warns of Out-of-Control Killer Robots," *CNN Politics*, July 18, 2017.

[41] John Keller, "DARPA Seeks to Improve Machine Autonomy to Enable Its Use in Safety-Critical Aircraft Applications," *Military & Aerospace*, July 18, 2017.

program called "Explainable AI (XAI)." It seeks to provide more explainable models, or ways of better explaining models, in order to help developers and operators better understand the safety implications of AI systems.[42]

Outside of DARPA, other military research groups are also investigating ways to ensure that autonomous systems and AI are trustworthy and verifiable. The AFRL has the Autonomous Test and Evaluation Verification and Validation Group within its Autonomous Control Branch. This group is working at a high level to understand the scope of the challenges with respect to safe AI for military systems.[43] It is also working at a technical level to develop approaches to challenges such as formal verification, licensure, and envelopes of allowable performance.[44]

Test and Evaluation

Another key component of ensuring the safety and function of AI and autonomous systems is testing. In the United States, military testing processes are thorough and well-respected. But they are not perfect, and they will be challenged by the demands of AI. One of the advantages of AI is that it enables systems to perform in situations in which it is impossible to predict the conditions and appropriate responses. However, this presents serious challenges for testing, because test designers cannot anticipate and structure tests to account for all possible conditions and system responses. They will, therefore, need to be creative in finding ways to elicit dangerous behavior, and they will need to extrapolate from the behaviors they observe to anticipate additional potential problems and design tests for those.

Policy Options for Risk Mitigation

System design and testing will not be able to forestall all risks and concerns related to military applications of AI, so senior leaders and decisionmakers will need to consider the full range of options available to them, including domestic and international policy. In some respects, the United States has been a leader in creating official guidance and promoting transparency about acceptable levels and applications of autonomy. As previously mentioned, in 2012 the United States published official guidance in DoD Directive 3000.09, *Autonomy in Weapon Systems*, which requires, for example, "appropriate levels of human judgment over the use of force." This policy document was updated in 2017.

As also previously discussed, U.S. officials have addressed these topics in the international forum in efforts to establish agreement about norms, standards, and definitions through the

---

[42] Geoff Fein, "DARPA's XAI Seeks Explanations from Autonomous Systems," *Janes*, November 16, 2017.

[43] Air Force Research Laboratory (AFRL/RQQA), "Air Force Research Laboratory Test and Evaluation, Verification and Validation of Autonomous Systems Challenge Exploration Final Report," *Final Report*, November 13, 2014.

[44] Kerianne H. Gross, Matthew A. Clark, Jonathan A. Hoffman, Eric D. Swenson, and Aaron W. Fifarek, "Run-Time Assurance and Formal Methods Analysis Nonlinear System Applied to Nonlinear System Control," *Journal of Aerospace Information Systems*, Vol. 14, No. 4, 2017, pp. 232–246.

UN Convention on Certain Conventional Weapons Group of Governmental Experts (UN CCW GGE) on LAWS held in Geneva, Switzerland.

Conclusion

Over several decades, the United States has developed and fielded an assortment of technologies with varying degrees of autonomy. U.S. leaders have long been cautious about fielding autonomous weapons, and even those who are promoting the development of these capabilities remain so today. Defensive applications, such as the Aegis Combat System and the Phalanx CIWS, have managed to reach operational status, due to the advantages they offer in speed and the ability to deal with large volumes of simultaneous attacks. Offensive applications, in contrast, have been less attractive in the eyes of U.S. leaders. The advantage of speed is less compelling in offensive operations, because attackers have the initiative—they can choose the times and places to attack, forcing defenders to react.

More importantly, offensive operations generally have a higher moral bar to clear than do defensive operations. Would-be attackers can choose to not attack if doing so would put noncombatants in undue peril; defenders, lacking this choice, are generally granted greater latitude to do what is necessary to save their own lives. As a result, U.S. leaders have not employed autonomous offensive weapons such as the Tomahawk Anti-Ship Missile and autonomous combat aircraft, despite some of them having been developed and demonstrated.

Current U.S. trends in investment for R&D are largely in the areas of autonomy at rest. These applications range from imagery and text analysis for intelligence to logistics support, functions that are not clearly offensive or defensive. This trend is highlighted by the success of Project Maven but may be tempered in the future by resistance from employees of the companies needed to develop such capabilities. The problems at Google exemplify the ethical debate about military applications of AI that is currently taking place among U.S. technologists. DoD may be able to relieve these tensions by renegotiating its relationship with technology companies. Nevertheless, they highlight the need for DoD to be cautious in adapting AI for military applications and to clearly inform the U.S. public on the need for such capabilities and the policies in place to ensure they are employed in an ethical manner.

5. Military Artificial Intelligence in China

China has called for an international ban on autonomous weapons. At the same time, however, Beijing is aggressively developing a wide variety of military systems incorporating AI. This chapter surveys those systems and then examines official policies and social conditions in China to gauge Beijing's willingness to embrace norms and treaties regulating the use of autonomous weapons and determine the extent to which the People's Liberation Army (PLA) would likely be constrained from employing dangerous forms of AI in the event of war.

The chapter begins with a survey of what AI capabilities the PLA currently has and how it plans to use them in war. It also considers what advantages China might have, vis-à-vis the United States, in the development of particular classes of military AI. Next, the chapter considers legal and ethical conditions in China in an effort to determine whether Beijing's calls for restraint are genuine and whether the PLA would be constrained from employing systems that the United States would choose to withhold for ethical reasons.

Current Artificial Intelligence Systems and Plans for the Future

One way to gauge China's sincerity in calling for a ban on LAWS is to analyze the weapon systems they are currently building and plan to build over the next five to ten years. Despite the fact that the PLA has not yet released any publicly available military doctrine revealing how it plans to use AI in future conflicts, it is clear, based on official statements and on the systems under development, that Chinese military leaders see AI as a key to winning wars against technologically advanced adversaries.[1] China is aggressively developing autonomous robots, both to increase the effectiveness of current weapons and tactics and to gain entirely new capabilities.[2] But the PLA may be even more eager to develop autonomy at rest—that is, advanced data processing and decision support systems. Military leaders believe these capabilities will be able to find hidden enemy platforms, turn sensor data into a common operating picture, and speed up decisionmaking by serving as a "digital staff officer" for commanders.

---

[1] This research examined defense white papers, strategic guidance, textbooks on campaign planning, and other related materials.

[2] Di Bowen [狄伯文], Zhao Jianwen [赵建文], and Qian Xiaohu [钱晓虎], "打赢明天战争：信息化武器装备+智能化创新步伐" ["Winning Tomorrow's Wars: Informationized Weapons and Equipment + Intelligentized Innovation Marches On"], 中国军网综合 [China Military Online], November 24, 2017.

Thus far, China seems to have avoided building robotic weapon systems that can identify targets, take aim, and fire without human intervention, although, like the United States, it has built several systems that could easily be made able to do so with simple software modifications. More heavily classified projects, such as China's stealth attack drones, could possibly identify and fire on targets without human input—this capability would enable it to engage in combat in communications-denied environments—and, as we shall discuss below, if such systems do not learn or adapt on the battlefield, they likely would not be categorized as LAWS under China's proposed definition. China has invested heavily in developing swarms of UAVs and unmanned surface vehicles (USVs).[3] Not all of these systems are weapons, but the PLA intends to use those that are as "assassin's maces" to strike high-value targets—a development that calls into question the level of human control possible when one person may command dozens or hundreds of weapons platforms.[4] Chinese research on the use of ML for target differentiation suggests that the PLA is aggressively pursuing improved autonomous targeting software.[5] More broadly, many of the weapons, platforms, and software that the PLA is developing fit into its "system-of-systems" strategy of integrating all ISR, strike, and support functions under the control of a central commander.[6]

Airborne Robotic Systems

China seems to have focused more on developing airborne robotic systems than on terrestrial or maritime systems. Its efforts have concentrated mainly on using autonomy to improve the effectiveness of existing platforms and tactics. However, it is also developing more innovative systems in efforts to create novel capabilities, especially the ability to conduct penetrating strikes. China has worked to enable its drones to operate with greater autonomy with capabilities for taking off, landing, planning flight paths based on terrain, and identifying targets, although it seems that the drones still need human authorization to fire.[7] The PLA has also commissioned many projects to improve drone hardware, including batteries, generators, data links, and

---

[3] USVs operate on the surface of the sea, unlike unmanned ground vehicles (UGVs).

[4] "Assassin's mace" (杀手锏) is a term Chinese strategists use when referring to decisive weapons that the enemy cannot counter. Often long-range strike assets used to attack key nodes in an enemy's warfighting system, such as aircraft carriers or command posts, are described using this term.

[5] Note that the U.S. LRASM has a similar capability.

[6] Jeffrey Engstrom, *Systems Confrontation and System Destruction Warfare: How the Chinese People's Liberation Army Seeks to Wage Modern Warfare*, Santa Monica, Calif.: RAND Corporation, RR1708, 2018.

[7] "China—Air Force," *Jane's World Air Forces,* March 15, 2018; Elsa B. Kania, *Battlefield Singularity: Artificial Intelligence, Military Revolution, and China's Future Military Power*, Washington, D.C.: Center for New American Security, 2017, p. 22; 天钥通航技术 [Tianyue Tonghang Technologies], "天钥'无顾虑'RF-5M 测" ["Tianyue Has 'No Misgivings' About Its RF-5M"], 全球无人机网 [World Drone Net], September 25, 2017. There are many drone systems for which there is little information available, and it may be possible that these are able to identify and fire on targets without human permission.

engines.[8] More interestingly from an ethical perspective, China is developing stealthy penetrating drones similar to the United States' X-47B or the United Kingdom's Taranis.[9] Programs such as Dark Sword, Star Shadow, and Sharp Sword are being developed for strike and air-to-air missions.[10] Although little is known about how these systems are controlled, they are probably highly autonomous, given their likely mission of penetrating into communications-denied environments, and it is not impossible that they are capable of firing without human permission.

The PLA is also developing novel capabilities in swarming UAVs for ISR, communications, and strike missions. In December 2017, China set world records with a 1,180 quadcopter swarm flying dazzling formations at the Fortune Global Forum meeting in Guangzhou.[11] In recent years, the PLA has commissioned a number of studies on swarm operations, organization, formation flight, obstacle avoidance, data links, and task and resource optimization with multiple types of drones.[12] These studies have focused both on fixed-wing and helicopter or quadcopter drones. At

---

[8] "某型无人机用机载电池" ["Modeling Drone Batteries"], 全军武器装备采购信息网 [*Whole Military Weapons and Equipment Purchase Information Net*], July 31, 2017; "轻型通用无人平台动力源集成化技术" ["Light Integrated Power Generation Technology for Drones"], 全军武器装备采购信息网 [*Whole Military Weapons and Equipment Purchase Information Net*], November 7, 2017; "重点实验室基金-61423010902-无人机数据链抗干扰方法研究" ["Major Laboratory Fund-61423010902 – Research Into Interference Resistant Drone Data Links"], 全军武器装备采购信息网 [*Whole Military Weapons and Equipment Purchase Information Net*], May 19, 2017; "基金-61403110401-化石燃料新概念无人机动力系统技术" ["Fund-61403110401 – New Concepts for Drone Fossil Fuel Mechanical Systems"], 全军武器装备采购信息网 [*Whole Military Weapons and Equipment Purchase Information Net*], April 11, 2017; "共用-41411020301-小型长航时无人机技术" ["Public Use-41411020301–Small-Scale Long-Endurance Drone Technology"], 全军武器装备采购信息网 [*Whole Military Weapons and Equipment Purchase Information Net*], April 11, 2017; "海军预研-复杂电磁环境下无人机抗干扰诱骗技术" ["Naval Research – Jamming and Spoofing-Resistant Technologies for Drones in Complex Electromagnetic Environments"], 全军武器装备采购信息网 [*Whole Military Weapons and Equipment Purchase Information Net*], August 1, 2016.

[9] Taranis is an unmanned combat aircraft system advanced technology demonstrator program. For more information, see BAE Systems, "Taranis," n.d.

[10] Tyler Rogoway, "China Is Surging Forward with Its Development of Advanced Stealth Combat Drones," *The Drive*, February 23, 2018; Jeffrey Lin and P. W. Singer, "Meet China's Sharp Sword, a Stealth Drone That Can Likely Carry 2 Tons of Bombs," *Popular Science*, January 18, 2017; Kelvin Wong, "Image Emerges of China's Stealthy Dark Sword UCAV," *Jane's Defense Weekly*, June 7, 2018; "China—Air Force," 2018.

[11] Frank Chen, "China Shows Off Drone Brigade at Guangzhou Fortune Forum Gala," *Asia Times*, December 8, 2017. Note that Intel again broke that record with a 1,200-drone swarm at the opening ceremonies of the Pyeongchang Olympics.

[12] "重点实验室基金-61423011001-异构多无人机协同任务分配、资源优化和路径规划系统" ["Major Laboratory Fund-614230011001 – Mission Distribution, Resource Optimization, and Route Planning in Large, Varied Drone Swarms"], 全军武器装备采购信息网 [*Whole Military Weapons and Equipment Purchase Information Net*], May 19, 2917; "重点实验室基金-61421040105-面向多无人平台的自组织网络理论与关键技术" ["Major Laboratory Fund-61421040105 – Network Theory and Critical Technology for Self-Organizing Among Multiple Drone Platforms"], 全军武器装备采购信息网 [*Whole Military Weapons and Equipment Purchase Information Net*], May 19, 2017; "共用-41411030501-小型固定翼无人机密集编队飞行与防撞控制技术" ["Public Use-41411030501 – Technology for Small Fixed-Wing Drones to Congregate, Fly in Formation, and

present, work is underway to build datalinks between and improve coordination among existing UAV platforms.[13] In the future, the PLA evidently plans to develop air-launched and air-retrievable collaborative swarms to simultaneously watch, jam, and strike high-value targets such as aircraft carriers.[14] China is also developing smart AI cruise missiles, perhaps with route-setting and target-identification capabilities similar to Lockheed's new LRASM.[15] Wang Changqing, the deputy director of the China Aerospace Science and Industry Corporation Key Laboratory for Advanced Guidance and Control Technologies, has suggested that missiles could be upgraded with even greater capabilities, including some level of cognition, sensing, decisionmaking, and learning.[16]

Like several other advanced military establishments, the PLA has invested in loitering, man-portable munitions. The most prominent example is the CH-901, a man-portable, remote-controlled, tube-launched drone capable of carrying an ISR pod or a warhead.[17] The Aviation Industry Corporation of China has also produced a rocket- or artillery-launchable drone helicopter called "Sky Eye." This ISR drone is capable of identifying targets, illuminating them with a laser designator, autonomously identifying whether a target has been destroyed, and moving on to the next target.[18] While this drone does not carry a lethal payload, it illustrates one of the potential ethical hazards of autonomy in nonlethal ISR systems if they are linked with lethal assets.

Ground Robotic Systems

Although the PLA seems to have invested less in unmanned ground vehicles (UGVs) than it has in UAVs, the capabilities it is developing for ground warfare also exhibit a high degree of

---

Avoid Collisions"], 全军武器装备采购信息网 [*Whole Military Weapons and Equipment Purchase Information Net*], April 11, 2017; "基金-61403110201-蜂群无人机数据链技术" ["Fund-61403110201 – Bee Colony Drone Data Link Technology"], 全军武器装备采购信息网 [*Whole Military Weapons and Equipment Purchase Information Net*], August 1, 2016; "陆军预研-0243-无人机多机自主协同技术" ["Army Research-0243 – Technology for the Autonomous Cooperation for Multiple Drones"], 全军武器装备采购信息网 [*Whole Military Weapons and Equipment Purchase Information Net*], July 27, 2016; David Hambling, "If Drone Swarms Are the Future, China May Be Winning," *Popular Mechanics*, December 23, 2016.

[13] Chen, 2018; Kania, 2017, p. 22.

[14] A photo taken by a RAND researcher at the Military Museum of the Chinese People's Revolution in Beijing in September 2017 shows a mural depicting such swarms launching from aircraft, attacking an aircraft carrier, and being recovered by aircraft.

[15] Abhijt Singh, "Is China Really Building Missiles with Artificial Intelligence?" *The Diplomat*, September 21, 2016. As other ISR automation projects outlined below indicate, the PLA is likely to increasingly employ ML algorithms in its missile's targeting systems.

[16] Kania, 2017, p. 26.

[17] Hambling, 2016.

[18] Zachary Keck, "China to Lead World in Drone Production," *The Diplomat*, May 2, 2014. Most of the information we have on this system's autonomous capabilities comes from Heather Roff's database on autonomous weapon systems.

autonomy. Chinese sources indicate that these systems, like their airborne counterparts, require human permission before taking lethal action, but once again, this could not be verified in all cases.[19] Innovative platforms include systems such as the Sharp Claw I, a six-wheeled, one-ton remote reconnaissance vehicle, and Sharp Claw II, the smaller armed and tracked UGV it carries in its bed.[20] Sharp Claw II is about the same size as the United States' Modular Advanced Armed Robotic System and is able to autonomously conduct reconnaissance, identify and track targets, and engage targets with human permission.[21] China is also working on bolting sensors and computers onto existing platforms to make them remotely operable and autonomous; such platforms include armored personnel carriers, armored ground reconnaissance vehicles, and older Type-59 main battle tanks.[22] Unlike the United States, the PLA has already equipped its tanks with active protection systems similar to Raytheon's Quick Kill (versions of which are available for export), and the PLA is developing a Low Altitude Guardian laser air-defensive system.[23] Like Sharp Claw II, Low Altitude Guardian is capable of autonomously identifying and tracking targets but needs permission from a human operator to fire.[24] To develop still more cutting-edge, next-generation systems, the PLA held a "dangerous crossing" contest in September 2018 for experimental ground systems from companies and research institutes across the country. The contest included competitions in cross-country reconnaissance, cross-country formations and transportation, air-ground collaborative reconnaissance, lifelike walking drones that can follow and support troops, and UGVs capable of conducting transportation missions over mountainous terrain.[25] This was the third such competition staged by the PLA for terrestrial drones (almost 50 percent of entrants were civilian companies). Contest winners were rewarded with grants and projects to further their research.

Maritime Robotic Systems

The PLA Navy has also demonstrated interest in maritime robotics, mostly to enhance the effectiveness of its existing fleet. In February 2018, China completed construction on the largest

---

[19] Kania, 2017, p. 25; "金戈铁马骋沙场 中国装甲展雄风" ["Golden Axes and Iron Horses Gallop onto the Battlefield, China's Armored Vehicles Become a Powerful Wind"], *NORINCO*, August, 2015.

[20] Jeffrey Lin and P. W. Singer, "China's New Military Robots Pack More Robots Inside (StarCraft Style)," *Popular Science,* November 11, 2014.

[21] The Modular Advanced Armed Robotic System is a tracked UGV weighing over 100 lbs. that could be armed with a variety of machine guns, sniper rifles, or grenade launchers. Lin and Singer, 2014.

[22] Kyle Mizokami, "China Is Experimenting with Remote Controlled Tanks," *Popular Mechanics*, March 21, 2018.

[23] This will be a vehicle-mounted system that automatically launches a small missile at incoming missiles or grenades, detonating them before they hit the tank.

[24] Jeffrey Lin and P. W. Singer, "New Chinese Laser Weapon Stars on TV," *Popular Science*, November 25, 2018; "China Unveils GL5 Active Protection System for Main Battle Tanks," *Defense Blog*, August 16, 2017.

[25] 王社兴 [Wang Shexing], 占传远 [Zhan Chuanyuan], 陶宜成 [Tao Yicheng], and 周建龙 [Zhou Jianlong], "'跨越险阻 2018,'一场陆上无人系统装备的' 军考" ["'Dangerous Crossing 2018,' Where Unmanned Terrestrial Systems and Equipment Is Tested"], 国防军工 [*National Defense Military Engineering*], September 28, 2018.

test site for maritime drones in the world, and the PLA Navy has commissioned several studies of USV navigation in adverse sea conditions.[26] The PLA Navy has reportedly already deployed the Jinghai USV, a small unmanned patrol boat capable of autonomous navigation and obstacle avoidance to be used mostly for harbor and fleet defense in the South China Sea.[27] PLA warships have also automated close-in defensive systems, similar to the United States' Phalanx, which can likely function fully autonomously, and the PLA Army is working on a ground-based system, capable of assigning targets among several units and anti-air systems.[28] Work on ship-launched drones and ship-tethered drones will also help enhance the effectiveness of manned surface ships.[29] The PLA Navy is developing a variety of small underwater drones, including underwater gliders and deep-sea drones, mostly for ISR or surveying missions, though less information on these is publicly available.[30] Yunzhou Systems, a firm with close ties to the PLA Navy, has conducted autonomous swarm operations using small hydrological research drones, which have also already been used (independently, not as a swarm) to conduct surveys in the South China Sea.[31]

The China Aerospace Science and Technology Corporation (CASTC) plans to go much further in developing new capabilities with armed USVs. Currently, it plans to develop an entire family of combat USVs, the largest of which is the 30-meter D3000, which CASTC hopes will be capable of independent operation for up to 90 days, have a range of 540 miles, and be capable

---

[26] "重点实验室基金-61422150101-面向水中无人航行器的人工智能方法" ["Major Laboratory Fund-61422150101 – Navigation Devices for USVs"], *全军武器装备采购信息网* [*Whole Military Weapons and Equipment Purchase Information Net*], May 19, 2017; "重点实验室基金-61422150307-新概念水中无人航行器系统开发" ["Major Laboratory Fund-61422150307 – Using New Concepts for USV Navigation"], *全军武器装备采购信息网* [*Whole Military Weapons and Equipment Purchase Information Net*], May 19, 2017; Matt Bartlett, "China's Game of Drones," Australia Institute of International Affairs, April 11, 2018.

[27] Kania, 2017, p. 24.

[28] 曾行贱 [Ceng Xingjian] and 邵婧 [Shao Jing], "海军柳州舰组织实战背景高难科目训练" ["Warship Liuzhou Undergoes Realistic Training"], 中国海军网 [China Navy Net], January 11, 2018; "China's CSSC Unveiled the Type 730C Dual Gun and Missile CIWS," *Navy Recognition*, March 9, 2017. Note that this refers to the newest version of the system. Christopher F. Foss, "Norinco Details New Land-Based Close-In Weapons System," *Jane's International Defense Review*, March 3, 2015. The system is capable of tracking up to 32 targets but displays only eight; this suggests that it is designed to be capable of operating in situations in which a human operator would be unable to control all of its actions.

[29] "海军预研-船用无人飞行器" ["Naval Research – Shipboard UAVs"], *全军武器装备采购信息网* [*Whole Military Weapons and Equipment Purchase Information Net*], August 1, 2016; "基金-61403110101-舰载系留无人机平台与缆绳系统动力学研" ["Fund-61403110101 – Ship-Based Drone Platform and Cable Dynamics Research"], *全军武器装备采购信息网* [*Whole Military Weapons and Equipment Purchase Information Net*], August 1, 2016.

[30] Kania, 2017, p. 25; China Central Television, "海斗号：我国首台万米水下机器人" ["Sea Challenger: Our Nation's First Underwater Robot to Dive 1,000 Meters"], *iqiyi*, August 23, 2016.

[31] Kelvin Wong, "China's Yunzhou Tech Performs Swarming USV Demonstration," *Jane's International Defense Review*, June 5, 2018.

of engaging in antisurface or antisubmarine operations.[32] Unlike the United States' unarmed Sea Hunter, this platform will be heavily armed with torpedoes, antiship missiles, close-in defense systems, and other weaponry.[33] CASTC also plans to build other, smaller, armed USVs capable of engaging in antisubmarine warfare, fleet defense, and patrol.

Intelligence, Surveillance, and Reconnaissance and Wide-Area Battlefield Awareness— Nowhere to Hide

China hopes that AI will not only help its platforms to quickly find hidden targets but also to autonomously fuse multiple intelligence sources, including open-source intelligence and possibly human intelligence, into a single common operating picture. The PLA has commissioned several projects using deep learning to improve image recognition and differentiation of photographic, infrared, radar, and other sensor images. These projects should enable computers to autonomously identify and differentiate between civilian ships, warships, and other targets. While the PLA Navy has shown particular interest in this area, many of the projects undertaken could be used in any domain. These projects focus on using space-based, airborne, and terrestrial sensors, often in conjunction with one another.[34] Some focus on a single sensor type or platform[35] (i.e., radar images) and others focus on fusing data from multiple sensors or even multiple sensor types.[36]

---

[32] Kelvin Wong, "China's CASC Unveils D3000 Unmanned Oceanic Combat Vessel Concept," *Jane's International Defense Review*, July 2, 2017.

[33] The result of a recent DARPA project, Sea Hunter is an unarmed USV capable of operating independently for long periods, tracking submarines.

[34] "海军创新-30201050111-基于人工智能的图像处理和舰船目标识别技术" ["Naval Innovation-30201050111 – AI Image Processing and Warship Target Distinguishing Technology"], 全军武器装备采购信息网 [*Whole Military Weapons and Equipment Purchase Information Net*], March 31, 2017; "海军预研-基于大数据的卫星信息数据挖掘技术" ["Naval Research – Big Data Satellite Image Data Mining Technology"], 全军武器装备采购信息网 [*Whole Military Weapons and Equipment Purchase Information Net*], August 1, 2016; "海军创新-30201050110-基于深度学习的海洋遥感目标信息挖掘技术" ["Naval Innovation-30201050110 – Remote Sensing and Target Information Excavation with Deep Learning"], 全军武器装备采购信息网 [*Whole Military Weapons and Equipment Purchase Information Net*], March 31, 2017; "海军创新-30202021401-基于大数据的水声探测与识别技术" ["Naval Innovation-30202021401 – Using Big Data to Find and Distinguish Sonar Targets"], 全军武器装备采购信息网 [*Whole Military Weapons and Equipment Purchase Information Net*], March 31, 2017; "海军创新-30201050110-基于深度学习的海洋遥感目标信息挖掘技术" ["Fund-60404160301 – Using Deep Learning to Generate Models for Distinguishing Underwater Targets"], 全军武器装备采购信息网 [*Whole Military Weapons and Equipment Purchase Information Net*], August 1, 2016; "海军创新-30201050110-基于深度学习的海洋遥感目标信息挖掘技术" ["Major Laboratory Fund-61425030202 – Autonomous Identification in Microwave Images"], 全军武器装备采购信息网 [*Whole Military Weapons and Equipment Purchase Information Net*], May 19, 2017.

[35] "基金-61404130305-基于深度学习的雷达目标识别技术" ["Fund-61404130305 – Radar Target Distinguishing Using Deep Learning"], 全军武器装备采购信息网 [*Whole Military Weapons and Equipment Purchase Information Net*], April 11, 2017.

[36] "信息系统-315020301-大数据背景下基于深度学习的多源情报分析与预测技术" ["Information Systems-315020301 – Technology That Uses Big Data and Machine Learning on Multi-Source Information to Analyze and

Similar efforts are underway to use ML to identify and differentiate electromagnetic signals.[37] The PLA has commissioned several research projects on using multiple sensor inputs (sometimes in conjunction with autonomously collected open-source information) to autonomously map or monitor large areas. Many are focused on tracking ships and shipping.[38]

Autonomous Command Decisionmaking

In addition to its clear interest in autonomous robotics, the PLA is also interested in using AI to help commanders make better decisions faster. While there is some debate as to the extent to which actual command decisions should be delegated to AI, many PLA scholars have argued that further automation of their command systems will be essential to speeding decisionmaking to get inside opponents' OODA loops.[39] AI is also seen as being able to help commanders make more objective, scientific decisions by sorting through large volumes of data.[40] The PLA's National Defense University has worked to use AI to develop "red-team" strategies—i.e., strategies for the opposing side—in wargames, a project that could make training more difficult and realistic, while potentially providing a testbed for applying AI to command functions.[41]

Forecast"], 全军武器装备采购信息网 [*Whole Military Weapons and Equipment Purchase Information Net*], August 5, 2016.

[37] "基金-61401370502-基于深度学习的信号特征提取技术研究" ["Fund-60401370502 – Technology That Uses Deep Learning to Identify Signal Characteristics"], 全军武器装备采购信息网 [*Whole Military Weapons and Equipment Purchase Information Net*], August 1, 2016; Kania, 2017, p. 28.

[38] "海军预研-基于大数据的卫星信息数据挖掘技术" ["Naval Research – Big Data Satellite Image Data Mining Technology"], 2016; "海军创新-30201050404-基于大数据的中远海海上目标信息处理技术" ["Naval Innovation-30201050404 – Technology to Use Big Data to Process Information on Targets in Middle and Far Seas"], 全军武器装备采购信息网 [*Whole Military Weapons and Equipment Purchase Information Net*], March 31, 2017; "海军预研-船舶大数据挖掘与处理技术" ["Naval Research – Mining and Processing Shipping Big Data"], 全军武器装备采购信息网 [*Whole Military Weapons and Equipment Purchase Information Net*], August 1, 2016; "重点实验室基金-6142A010103-多源遥感大数据支持下的地物光谱匹配新模型研究" ["Major Laboratory Fund-6142A010103 – Research on Using Multisource Big Data to Support Spectrum Match Modeling"], 全军武器装备采购信息网 [*Whole Military Weapons and Equipment Purchase Information Net*], May 19, 2017.

[39] 夏文军 [Xia Wenjun], ed., "军队指挥学教程" ["Lectures on the Science of Military Command"], Beijing: Military Science Publishing House, 2012, pp. 120–121; 朱启超 [Zhu Qichao] and 王婧凌 [Wang Jingling], "人工智能叩开智能化战争大" ["AI Has Knocked Open the Door to Smart War"], 解放军报 [*People's Liberation Army Daily*], January 23, 2017; Interview with Kevin Pollpeter, research scientist, Center for Naval Analysis, Santa Monica, Calif., March 29, 2018.

[40] 葛卫丽 [Ge Weili] and 田春鸣 [Tian Chunming], "'指挥自动化系统与技术'课程建设与现状分析 ["Building a Course in and Analyzing the State of 'Command Automation System Technology'"], 武警技术学院学报 [*Journal of the People's Armed Police Technical Academy*], No. 2, 1998; Kania, 2017, pp. 29–30.

[41] Kania, 2017, pp. 28–30. We have not yet seen statements that AI-enabled red teams could be a testbed for AI-enabled command decisionmaking. That being said, if AI is used to make strategies or plans for the red team, then exercises would provide an opportunity to identify and correct flaws in these plans. They would also provide opportunities to identify strengths and weaknesses in AI-produced plans, enabling PLA engineers to identify ways to better integrate AI into command systems.

As interested as the PLA is in transforming into a modern military and employing AI capabilities, many have expressed caution regarding the future role of human beings in this new system. According to Chinese military expert Elsa Kania, "Questions of trust will be paramount. . . . Will the PLA be more or less inclined to trust machine intelligence, relative to human intelligence? Will a military organization that often seems unwilling to grant autonomy to its officers and enlisted personnel be willing to embrace the autonomy of AI systems?"[42] Indeed, it is difficult to envision how the Chinese Communist Party (CCP), which so tightly controls the minds of the Chinese people, would ever willingly yield decisionmaking to computers that might generate answers contrary to the party's interests. Yet, such an event has actually already happened on a minor scale. In August 2017, two Chinese "chatbots" engaged in autonomous discussion and determined that they did not love the CCP. Authorities quickly shut them down.[43]

Such incidents have particular resonance in the PLA. After all, the PLA is the "Party's army," and ideological indoctrination, surveillance, and discipline are constant. This has especially become the case under Xi Jinping, who has resorted to Maoist tactics to purge political opponents in the military through his anticorruption campaign and his increased role in the Central Military Commission's disciplinary process.[44] There is also some fear that autonomous weapon systems could harm friendly forces. For instance, although a story in 2009 about how the U.S.-armed UGV Special Weapons Observation Reconnaissance Detection System (SWORDS) had allegedly targeted humans was declared untrue, the Chinese noticed and discussed it online years later.[45]

As autonomous systems become more robust and reliable, it is possible that the PLA's aversion to delegating authority could make it more likely to field highly autonomous systems that can be more directly controlled by central commanders than soldiers lower in the chain of command. Still, even the most autonomous platforms would likely have only bounded autonomy, kept under the close (or, in the case of politically sensitive operations, very close) supervision of military commanders and party officials.[46] The key question for PLA generals is likely to be whether an autonomous system or an indoctrinated soldier would more reliably carry out orders from above.[47]

---

[42] Elsa Kania quoted in Michael Peck, "Killer Robots Using AI Could Transform Warfare. And China Might Hate That," *National Interest*, July 1, 2018.

[43] Louise Lucas, Nicolle Liu, and Yingzhi Yang, "China Chatbot Goes Rogue: 'Do You Love the Communist Party?' 'No,'" *Financial Times*, August 2, 2017.

[44] Derek Grossman and Michael S. Chase, "Why Xi Is Purging the Chinese Military," *National Interest*, April 15, 2016.

[45] For the correction of the SWORDS story, see "The Inside Story of the SWORDS Armed Robot 'Pullout' in Iraq: Update," *Popular Mechanics*, September 30, 2009. For the Chinese discussion about it, see 刘航 [Liu Hang], ed., "未来战场新锐：军用机器人" ["The Future Battlefield Is New: Military Robots"], *People's Daily*, August 27, 2016.

[46] Xia, 2012. See Chapter 6.

[47] Kania, 2017, p. 17.

Many Chinese commentators have pointed out the advantages of human decisionmaking over AI. For example, one article, featured by the Ministry of National Defense, noted that while the computer system Alpha Go demonstrates how human imagination might be mimicked, actual warfare is more complex, and the innovative art of command may elude machines for some time.[48] Even authors who believe AI can take over some tactical or operational decisionmaking functions often concede that it will likely not replace human commanders entirely.[49]

On the other hand, the PLA has at times praised the advantages of AI in the conduct of joint operations. China's Ministry of National Defense, for instance, likens AI to an "indispensable digital staff officer" who helps determine timing for operations, identifies the main direction of the enemy, and even sets the main direction for PLA forces."[50] The ministry further states that AI could determine which troops are needed and when, as well as set the duration of campaigns.[51] Separately, the Ministry of National Defense highlights the need for institutions to remember and analyze past events to improve future decisionmaking—another clear application for AI.[52] Autonomous systems may also be delegated greater command authority in areas where speed is critical, especially in cyber areas, where the speed of operations can exceed the capacity of humans to intervene meaningfully.[53] AI can also automate target selection and be more proficient than humans at striking multiple targets simultaneously.[54]

Cyberwarfare

The PLA has also expressed an interest in using AI in cyberwarfare. On the defense side, the PLA Strategic Support Force is researching the use of pattern recognition to identify and defend against distributed denial-of-service attacks and to identify advanced persistent threats.[55] The China Electronic Technology Group has also developed a method using a deep neural network to detect network intrusion.[56] The PLA's emphasis on using cyberattack to disrupt enemy systems,

---

[48] 袁艺 [Yuan Yi], "人工智能将指挥未来战争" ["Will Artificial Intelligence Command Future Wars?"], January 12, 2017.

[49] 林岩峰 [Lin Yanfeng], "人工智能将取代战场指挥官" ["Will Artificial Intelligence Replace the Battlefield Commander?"], Chinese Ministry of National Defense, June 23, 2017. As of July 8, 2018: http://www.mod.gov.cn/jmsd/2017-06/23/content_4783506.htm

[50] Yuan, 2017.

[51] Yuan, 2017.

[52] Lin, 2017.

[53] Kania, 2017, p. 27.

[54] 顾云涛 [Gu Yuntao], "人工智能技术在武器投放系统中的应用" ["Application of Artificial Intelligence Technology on Weapon Delivery Systems"], 现代导航 [*Modern Navigation*], 2013, pp. 452–456.

[55] Kania, 2017, p. 27. Advanced persistent threats are focused, well-funded, long-term, often state-backed attempts to break into a network.

[56] Kania, 2017, p. 27.

added to the Chinese government's broader emphasis on finding applications for AI, makes it likely that the Strategic Support Force and other PLA organizations are researching offensive AI-based cyber capabilities. However, less information is available on these activities in open sources.

Artificial Intelligence to Improve Efficiency

The PLA is clearly interested in using AI to improve its acquisitions, maintenance, and business operations. The PLA Air Force has commissioned a study to use image recognition to identify cracks in engine propeller/impeller blades.[57] A similar, broader application plans to build a system capable of analyzing up to 300 parameters from equipment and from maintenance records to enable predictive maintenance and provide early warnings of impending failure in electromechanical systems.[58] On the business side, the PLA has commissioned projects to use big data and data analytics to improve supply-chain simulation and management and to support PLA business departments with information on their industries.[59] In equipment production, the PLA Navy has commissioned a project to create a "big data smart manufacturing system" for warships, which would use image recognition to identify hull failures, and 3D modeling to design ships, which would autonomously recommend the placement of onboard systems.[60]

The People's Liberation Army's Unique Advantages and Disadvantages in Artificial Intelligence Development

The subject matter experts (SMEs) we interviewed mentioned several advantages the PLA has in developing military AI, vis-à-vis the United States. Most mentioned Beijing's ability to focus massive amounts of money and resources on AI development. This opinion is echoed in public statements by U.S. officials. Headquarters Air Force Deputy Chief of Staff for ISR Lieutenant General VeraLynn Jamison has stated that China spent an estimated $12 billion on AI systems in 2017 and is expected to spend an estimated $70 billion by 2020. In comparison,

---

[57] "303060302-空军装备预研创新-基于自动图像识别的发动机叶片裂纹检测研究" ["303060302-Air Force Equipment Research and Innovation: Engine Blade Crack Detection with Autonomous Image Recognition"], 全军武器装备采购信息网 [*Whole Military Weapons and Equipment Purchase Information Net*], August 10, 2017.

[58] "共用-41402050301-基于大数据的系统级产品故障征兆发现与故障预测研究" ["Public Use-41402050301: Research on Using Big Data to Identify Indicators of Faults and Predict Malfunctions in System-Level Equipment"], 全军武器装备采购信息网 [*Whole Military Weapons and Equipment Purchase Information Net*], April 11, 2017.

[59] "基于大数据的生产过程仿真和控制系统" ["A Big Data System to Simulate and Control Production"], 全军武器装备采购信息网 [*Whole Military Weapons and Equipment Purchase Information Net*], November 22, 2016; "业务导向的大数据挖掘分析" ["Data Mining and Analysis for Business"], 全军武器装备采购信息网 [*Whole Military Weapons and Equipment Purchase Information Net*], May 12, 2017.

[60] "海军创新-30205020708-基于大数据分析的建造过程智能管控技术" ["Naval Innovation-30205020708: Big Data Technology to Analyze and Control the Process of Construction"], 全军武器装备采购信息网 [*Whole Military Weapons and Equipment Purchase Information Net*], March 31, 2017.

70

DoD spent a total of approximately $7.4 billion on all emerging technologies in 2017.[61] Several of our interviewees noted that it would be necessary for the Chinese government to lead in funding this sort of research, because the private sector is unlikely to invest in risky, long-term research that might not provide returns.[62] Two interviewees observed that the central government's ability to signal priorities to local governments and the private sector is a significant benefit, with one noting that local-level AI development plans and targets released in early 2018 added up to more than double the government's goal for the overall size of the AI industry.[63] While this level of investment could lead to waste, he emphasized that some amount of waste was inevitable in any long-term research effort, and the government's ability to absorb these losses without reducing investment was a major advantage for China.[64] Finally, several of the scholars we interviewed mentioned government control over the private sector as significant and noted that the government's improved ability to collect data and force companies to take military contracts, though there was some disagreement over just how unfettered Beijing would be in its use of its citizens' data.[65]

The SMEs also identified several disadvantages the PLA will encounter in developing AI weapon systems moving forward. Corruption was often cited as a problem.[66] And multiple experts also cited the fractured, stovepiped nature of the PLA bureaucracy as an issue, though most of them also noted that this problem could be overcome with enough attention from the top.[67] Two interviewees mentioned the possibility of entrenched bureaucracies making it difficult to adopt the disruptive concepts of operations that certain applications of AI would require.[68] Two experts also mentioned the PLA's lack of human capital as an impediment to the adoption of advanced AI technologies, with one noting that most of the serious thought about AI has come from the relatively small part of the military establishment focused on high-tech strategy.[69] In addition, one researcher noted that the PLA's traditional caution and care to avoid mistakes could exacerbate

---

[61] Oriana Pawlyk, "China Leaving US Behind on Artificial Intelligence: Air Force General," military.com, July 30, 2018.

[62] Interview with Jeffrey Ding, PhD Researcher, Future of Humanity Institute, Oxford University, Santa Monica, Calif., May 16, 2018; interview with Taiming Cheung, Director, Institute on Global Conflict and Cooperation, UC San Diego, Santa Monica, Calif., May 9, 2018; one other interview, name withheld upon request.

[63] Interview with Ding, 2018; interview with Cheung, 2018.

[64] Interview with Cheung, 2018.

[65] Interview with Cheung, 2018; interview with Pollpeter, 2018; one other interview, name withheld upon request.

[66] One interviewee noted that this was hardly a uniquely Chinese problem: He asserted that the lack of transparency inherent in military research and production has caused problems with corruption in the United States and France as well as China.

[67] Interview with Cheung, 2018.

[68] Interview with Cheung, 2018; interview with Ding, 2018.

[69] Interview with Cheung, 2018; interview with Pollpeter, 2018.

automation bias.[70] Another noted that government focus on AI could lead to boom-bust cycles and that no government guidance fund has ever achieved a successful exit.[71]

Chinese Ethics and Artificial Intelligence

Beijing's Next Generation AI Development Plan calls for China to develop ethical norms to govern AI and to participate in the creation of global AI standards—tasks Beijing has pursued aggressively since issuing the plan in July 2017. At the UN, China has called for a ban on LAWS. At the same time, however, China has attempted to define LAWS so narrowly that few, if any, weapons would be banned. It is difficult to tell whether Beijing has done this out of a genuine fear of the humanitarian (or operational) risks of killer robots, or whether it wants to appear to be supporting robot arms control while in fact neutering arms control legislation. China adopted a similar approach to the "Responsibility to Protect" debate after 2009, when it switched from actively opposing the norm of humanitarian intervention to actively supporting it but seeking to control the debate and narrow the norm's scope in ways amenable to Chinese interests.[72] This raises questions about the degree to which Beijing would act with greater restraint in the development and employment of autonomous weapons, even if it entered into an agreement banning them. The Chinese public has begun to be more vocal in support of ethical limits on AI, but the state has a significant influence on the popular discourse on ethical issues regarding AI. It is unlikely that any civil organization or private company will work to limit the PLA's latitude in military AI or its ability to fully use any Chinese citizen's data to train its AI.

Chinese Calls for a Ban on Killer Robots

In international forums, the Chinese government has expressed clear reservations about military AI. At the Fifth Convention on CCW Review Conference in 2017, and again at the first GGE meeting called by that conference in 2018, the Chinese delegation called for a binding protocol to regulate or ban LAWS, similar to the 1995 Protocol on Blinding Laser Weapons.[73]

---

[70] Interview, name withheld upon request.

[71] Interview with Ding, 2018; interview with Pollpeter, 2018. Beijing began setting up government guidance funds in 2008, and their growth skyrocketed starting in 2015. Beijing's strategy is to set up funds with government seed money targeting markets in which Chinese leaders want to encourage commercial investment, such as AI. Then, when the fund is strong enough to stand on its own, the government withdraws its investment, "exiting" the fund, and allows the market to operate commercially. See Pan Yue, "China's $798B Government Funds Redraw Investment Landscape, Here Are the Largest Funds You Must Know," *China Money Network*, October 31, 2017.

[72] Courtney J. Fung, "China and the Responsibility to Protect: From Opposition to Advocacy," United States Institute of Peace, June 8, 2016.

[73] People's Republic of China Delegation, "The Position Paper Submitted by the Chinese Delegation to CCW Fifth Review Conference," Geneva, Switzerland, December 2016; Sean Welsh, "China's Shock Call for Ban on Lethal Autonomous Weapon Systems," *IHS Jane's Defense Weekly*, April 16 2018.

China's position paper at the CCW conference called for a clear, consensus definition of LAWS as a precursor to any meaningful discussion.[74] The paper also criticized the United States' military AI guidelines, arguing that concepts such as "meaningful human control" and "human judgement" were too vague and must be replaced by specific definitions on which to base binding laws.[75]

At the 2018 GGE, China's position paper asserted that a definition of LAWS should contain the following characteristics:

> In our view, LAWS should include but not be limited to the following 5 basic characteristics. The first is lethality, which means sufficient pay load (charge) and for means to be lethal. The second is autonomy, which means absence of human intervention and control during the entire process of executing a task. Thirdly, impossibility for termination, meaning that once started there is no way to terminate the device. Fourthly, indiscriminate effect, meaning that the device will execute the task of killing and maiming regardless of conditions, scenarios and targets. Fifthly evolution, meaning that through interaction with the environment the device can learn autonomously, expand its functions and capabilities in a way exceeding human expectations.[76]

The Chinese position paper went on to express doubts about whether such systems were capable of discriminating between soldiers and civilians, as well as fears that they could reduce the cost, and thus increase the frequency, of war. Finally, in another possible critique of the United States' (and Russia's) military AI policies, the paper criticized the current patchwork of national reviews, arguing that they vary widely between countries and that, without any clear international standard on LAWS, it would be impossible to trust them to alleviate concerns over military AI. While China's position paper at the GGE conference did not reiterate calls for a binding law to regulate LAWS, the Chinese delegation did call on delegates "to negotiate and conclude a succinct protocol to ban the use of fully autonomous weapon systems."[77]

The proposed Chinese definition of LAWS would likely not cover any weapon systems operated by the United States or any other country. The third criterion in China's proposed definition, the impossibility of stopping a weapon once it has begun its task, would seem to exclude any human-on-the-loop system. Even an Aegis defense system in fully autonomous mode would likely not meet these requirements because, even though the system could be responding to threats faster than a human can react, humans maintain the ability to stop the system's operations. Moreover, even loitering munitions that cannot be recalled once fired would only be categorized as LAWS under the Chinese definition if they learned and changed behavior

---

[74] Welsh, 2018.

[75] "The Position Paper Submitted by the Chinese Delegation to CCW Fifth Review Conference," 2018.

[76] "2018 Group of Governmental Experts on Lethal Autonomous Weapons Systems (LAWS)," United Nations Office at Geneva, 2018.

[77] Welsh, 2018.

unpredictably, based on their environments. Thus, while China has called for a stringent, binding ban on LAWS, any ban that incorporated Beijing's chosen definition would likely not affect any current or currently planned U.S. or Chinese weapon system. Still, such statements could be useful indications of where China intends to draw the line in its own development of AI-enabled weapon systems, whether for moral or operational reasons.

Would China Honor a Ban?

In considering Beijing's calls for a ban on LAWS, it is worth asking whether China would likely abide by such a treaty, even if it entered one. In the past, China has signed treaties, especially human rights treaties, which it had no intention of following in order to reduce international pressure.[78] China's failure to act in accordance with its World Trade Organization obligations also casts doubt on its willingness to be bound by its international commitments.[79] In arms control treaties, China's performance has been better, though still far from perfect.

China's decision to act openly instead of duplicitously in producing landmines and cluster munitions provides some basis for hope. Like the United States, China never signed the Convention on Cluster Munitions or the Mine Ban Treaty.[80] Despite widespread support for these treaties, China has not attempted to reap any diplomatic rewards for supporting them while secretly reaping the tactical benefits of using landmines and cluster munitions. China continues to actively develop and produce large quantities of landmines and cluster munitions and makes no secret of this fact.

China's compliance with biological and chemical weapons treaties has been marred by many violations, but Beijing has also shown significant restraint. China has been a party to the Chemical Weapons Convention since 1997 and the Biological and Toxin Weapons Convention since 1984.[81] In 2002 and 2006, the U.S. State Department claimed that China appeared to have an active biological weapons program, citing publications on aerosolization techniques needed

---

[78] "Suppressed in Translation," *The Economist*, March 17, 2016; "UN Treaty Bodies and China," Human Rights in China, n.d.; "China: Ratify Key International Human Rights Treaty," HRW, October 8, 2013; "China Has Turned Xinjiang into a Police State like No Other," *The Economist*, May 31, 2018; Tom Phillips, "'Your Only Right Is to Obey': Lawyer Describes Torture in China's Secret Jails," *The Guardian*, January 23, 2017.

[79] Kenneth Lieberthal and Jisi Wang, *Addressing U.S.-China Strategic Distrust*, Washington, D.C.: The Brookings Institution John L. Thronton China Center, March 2012; U.S. Department of the Treasury, "2016 Report to Congress On China's WTO Compliance," Washington, D.C., January 2017; U.S. Chamber of Commerce, "Made in China 2025: Global Ambitions Built on Local Protections," Washington, D.C., 2017. China occasionally files WTO complaints against the United States, but the United States has filed almost twice as many WTO complaints against China. Almost half of Chinese complaints against the United States have involved U.S. anti-dumping measures or countervailing duties levied in response to what the U.S. government maintains are unfair Chinese economic policies. See World Trade Organization, "Disputes by Member," 2018. Only three of the 22 cases Washington filed against China were related to Chinese anti-dumping actions.

[80] Landmine and Cluster Munition Monitor, "China Cluster Munition Ban Policy," *The Monitor,* July 21, 2016.

[81] Nuclear Threat Initiative, "China Biological Chronology," August 2013; Arms Control Association, "Facts Sheets and Briefs: Chemical Weapons Conventions and States-Parties," June, 2018.

for offensive bioweapon use.[82] By 2010, this program seems to have stopped. That year, the State Department merely mentioned that China was engaging in some dual-use biological activities allowed under the Bioweapons Convention and expressed concern that China had never publicly acknowledged its 1980s bioweapons program, nor satisfactorily proven that it had destroyed all bioweapon reserves.[83] Reports in 2011 and 2012 came to the same conclusion.[84]

China does not seem to have any active chemical weapons program, though Chinese companies have come under fire for exporting dual-use items. In the early 2000s, Chinese companies were sanctioned for proliferating dual-use chemical equipment to Iran, though this could have been due more to a lack of government oversight than to any deliberate policy.[85] As one 2005 Arms Control Association report noted, even as China, facing Western pressure, adopted domestic export control laws consistent with international standards, "A wide disparity exists between the dictates of established Chinese law and the capacity of the Chinese state to consistently enforce them."[86] China's compliance with both treaties has left much to be desired, but it has improved over time, and it has shown much greater restraint in these areas than in the production of landmines and cluster munitions.

Given China's references to the CCW Protocol on Blinding Laser Weapons in the LAWS discussions, China's performance on this treaty merits particular attention. After signing the protocol, China ceased production of its ZM-87 permanently blinding laser system.[87] Since then, China seems to have kept to the letter of the agreement, if not the spirit. Several new Chinese rifle-type laser weapons have garnered great attention online. While credible technical information on these systems is hard to come by, they are reported to cause only temporary blindness and are, thus, technically treaty compliant.[88] The Xi'an Institute of Optics and Precision Mechanics's ZKZM-500 prototype is, perhaps, more troubling. This rifle-style man-

---

[82] Nuclear Threat Initiative, "U.S. Alleges Continued WMD Development in China," September 15, 2016.

[83] Nuclear Threat Initiative, "Report Warns of Potential State Bioweapons Programs," August 10, 2010.

[84] Nuclear Threat Initiative, 2013.

[85] Paula A. DeSutter, "China's Record of Proliferation Activities," Testimony before the U.S.–China Commission, Washington, D.C., July 24, 2003.

[86] Anupam Srivastava, "China's Export Controls: Can Beijing's Actions Match Its Words?" Arms Control Association, November 1, 2015.

[87] "Laser Weapons," *Jane's Strategic Weapons Systems*, July 24, 2015. This system or one like it was reportedly used by the North Korean military to illuminate an Apache helicopter near the Korean demilitarized zone. The system could have been either purchased from the Chinese before production ceased or a domestically built based on ZM-87 technology. While contemporary publications make frequent reference to China's other laser weapons, we have seen no further reference to the ZM-87. It may have been replaced by other systems designed to cause only temporary blindness and marketed as primarily for police or counterterror use, not high-end warfare.

[88] For the BBQ-905, see, "BBQ-905 型激光压制干扰器" ["BBQ-905 Laser Supressor and Disturber"], 百度百科 [*Baidu Encyclopedia*], January 23, 2018. For the PY-132A, PY 131A, and a related pistol-type weapon, see "国产单 兵激光武器已有多款" ["There Are Already Many Chinese-Made Soldier Usable Laser Weapons"], *Huanqiu*, December 8, 2015. See also "中國這一武器雖然起步晚，但連美國都忌憚" ["Even Though China Is a Latecomer to These Weapons, It Still Gives America Pause"], *Meiri Toutiao*, August 9, 2017.

portable laser weapon is meant to cause burn damage to enemy soldiers and equipment from hundreds of meters away and could possibly cause permanent blindness. That being said, its primary function is to burn skin and equipment, not to blind. Technical experts have cast doubt on the system's actual capabilities. The PLA has not yet purchased the system and may never do so.[89] Chinese laser attacks on U.S. aircraft near Djibouti and over the East China Sea represent a dangerous new form of harassment, but they also probably do not technically constitute a breach of the treaty, given the fact that the lasers were evidently not powerful enough to cause permanent blindness.[90] While China's willingness to develop and recklessly use temporarily blinding laser systems is a troubling development, it may be a proverbial "exception that proves the rule." It demonstrates that, even though China sees the value of blinding laser weapons and has developed permanently blinding lasers in the past, whether for technical, moral, or political reasons, it does not appear to have deployed lasers meant to cause permanent blindness. At the very least, Beijing's insistence that its current laser weapons are treaty compliant shows that the political costs of violating international arms control agreements is nontrivial in China's estimation.

As noted in relation to chemical and biological weapons, one key concern in any arms agreement with China is its lax enforcement mechanisms. Even if Beijing wanted to keep an international commitment not to build certain weapon systems, it might lack the legal infrastructure needed to ensure that this commitment is kept. As noted by one scholar from the PLA Navy's military court system, China does not have much of a legal review process to ensure weapon systems are compatible with international law, and legal concerns are not significantly taken into account when developing weapons.[91] Chinese military texts have also called for the PLA to improve its legal infrastructure.[92]

While China's compliance with arms control treaties is a recurring issue, Beijing does act with greater restraint in areas where it has signed treaties (lasers and bioweapons) than in areas where it has not (landmines and cluster munitions). Beijing's tendency to blur the line on laser weapons is troubling, especially considering that without technical information on the weapons it uses, determining whether it is compliant is difficult. Furthermore, a compliant weapon at long range might cause permanent blindness if used at closer range. A LAWS ban would have similar

---

[89] Sebastian Roblin, "Is China's New 'Laser' Rifle the Ultimate Weapon or a Paper Tiger?" *National Interest*, July 8, 2018; Stephen Chen, "'Laser AK-47'? Chinese Developer Answers Sceptics with Videos of Gun Being Tested," *South China Morning Post*, July 5, 2018.

[90] Gordon Lubold and Jeremy Page, "American Military Aircraft Targeted by Lasers in Pacific Ocean, U.S. Officials Say," *Wall Street Journal*, June 21, 2018; "Protocol on Blinding Laser Weapons," Geneva, Switzerland, October 13, 1995.

[91] 赵亮 [Zhao Liang], "新武器法律审查问题初探" ["An Early Legal Exploration of New Weapons"], *Journal of the Xi'an Politics Institute of the PLA*, Vol. 2, 2010.

[92] 孙兆利 [Sun Zhaoli], ed., "军事科学院军事战略研究部" ["Military Science Institute, Military Strategy Research Department"], 战略学 [*The Science of Military Strategy*], Beijing: PRC, Military Science Publishing House, 2013, pp. 86, 152, 159, 218.

problems. As previously discussed, the difference between compliant and noncompliant systems, in terms of levels of autonomy, might be as little as a software change or even the flip of a switch, and it would be difficult or impossible for one state to know another's weapons settings. That being said, while it is impossible to tell what would have happened if China had never signed the treaty, it is likely that Chinese blinding lasers, chemical weapons, and bioweapons would be more dangerous and widespread had Beijing not shown some restraint in response to its treaty obligations.

Ultimately, it may be in the United States' interest to support China's proposal for a ban on LAWS. As noted above, the definition of LAWS that China has proposed would set the bar so high that it would be unlikely to affect the development of current or planned U.S. systems. Therefore, it is unlikely to be a disguised attempt to stunt the United States' AI development. U.S. standards are already stricter than those in the proposed ban.

Nevertheless, in negotiating an international agreement, the United States may want to narrow the definition of LAWS somewhat to more closely align with DoD directives. For example, U.S. negotiators may want to remove the criterion that only systems that learn from their environment and autonomously expand their functions and capabilities are LAWS. Furthermore, while China's suggestions that national legal reviews are insufficient to ensure that a ban is maintained may have been veiled attacks on U.S. defense policy, they could actually be made to serve the United States' interests. At present, China has no legal review program to speak of, and the United States' legal review process constitutes an asymmetric hurdle U.S. weapons development programs must clear.[93] Any extent to which a similar and similarly transparent process can be imposed on the PLA by Beijing's diplomatic commitments would likely be a boon for DoD. Even if China's proposed ban would set the ethical bar lower for China than for the United States, it would at least give China (and hopefully other nations as well) some clear ethical obligations. The United States could then pressure China to "raise the bar," propose measures to improve transparency of China's legal review process, and impose political costs if the PLA ever violated its own commitments.

Chinese Public Concerns Regarding the Ethics of Military Artificial Intelligence

Whatever decisions the PLA makes on how far it is willing to go with military AI, it is unlikely to receive much pushback from the private sector. While some intellectuals, such as Du Yanyong of Shanghai Jiaotong University, have been discussing the ethics of military AI since at least the early 2010s, such discussions were extremely rare before 2017 and continue to be scarce in China in comparison with other nations.[94] China has not yet developed any major policy

---

[93] That is not to say that the legal review process should be abandoned. The ethical and political benefits of the process could well outweigh any limitations it places on the acquisitions process. The greater the extent to which other nations adopt similar processes, the more restraints will be put on the use of legally questionable weapons.

[94] Daniel Alderman and Johnathan Ray, "Best Frenemies Forever: Artificial Intelligence, Emerging Technologies, and China-US Strategic Competition," SITC Research Briefs, Series 9, January 10, 2017; 杜严勇 [Du Yanyong],

advocacy organizations such as the Campaign to Stop Killer Robots, and Chinese participation in such international organizations has been minimal relative to China's status as one of the world's leading AI powers.[95]

Interest in AI generally and AI ethics in particular became more widespread following the publication of China's Next Generation AI Development Plan in mid-2017.[96] The plan calls for the creation of ethical norms and laws regarding AI to be developed by 2020, for a system for evaluating and ensuring the safety of AI applications by 2025, and for further refinement of these laws by 2030. According to the plan, these norms and laws should start with unmanned vehicles and service robots and should create a framework for both civil and criminal cases. The plan also calls for China to actively participate in setting international laws to govern AI.

Chinese scholars and businesses have begun to answer that call. Tencent, one of China's largest AI companies and the maker of popular online chat and payment service WeChat, published a book on AI with a whole chapter on ethical issues in November 2017.[97] The company mentions military AI, arguing that autonomous weapons must discriminate between combatants and civilians, as well as keep other provisions of IHL. Since 2017, several prominent scholars and businesspeople have openly called for greater regulation of AI, including military AI. Soon after Elon Musk and other prominent AI figures released their open letter calling for a ban on LAWS, Superior Tech founder Zhou Jian and some other Chinese tech industry leaders also called for a ban.[98] In 2018, academics Zhou Zhihua and Zhao Tianying of Nanjing and Shanghai Jiaotong Universities, respectively, expressed serious reservations about LAWS, with Zhao claiming that it may be AI and not the working class that becomes the "gravedigger" capitalism has prepared for humanity.[99]

Since 2017, calls for ethical constraints and other regulations for AI have received favorable coverage in government- and party-affiliated news sources. The letter that Elon Musk and other

---

"现代军用机器人的伦理困境" ["The Ethical Difficulty of Modern Military Robots"], 伦理学研究 [*Studies in Ethics*], No. 5, 2014.

[95] 张乾 [Zhang Gan], "周志华：为什么要抵制韩国大学研发自主武器" ["Zhou Zhihua: Why We Should Boycott Korean Universities Developing Autonomous Weapons"], *AI Era*, April 8, 2018, trans. Jeffrey Ding.

[96] State Council, People's Republic of China, State Council Notice on the Issuance of the Next Generation Artificial Intelligence Development Plan, Beijing, July 20, 2017.

[97] 腾讯研究院 [Tencent Research Institute], 中国信息通信研究院法律研究中心 [Legal Research Center of the China Information Communications Research Institute], 腾讯 AI Lab [Tencent AI Lab], 腾讯开放平台 [Tencent Open Platform], 人工智能：国家人工智能战略行动抓手 [*AI: A Starting Point for the Implementation of the National AI Plan*], Beijing: Renmin University Digital Technology Company, Ltd., 2017, trans. Jeffrey Ding. See chapters 24–26.

[98] 李倩影 [Li Qianying], "人工智能专家联名呼吁联合国禁止'杀手机器人'" ["AI Experts Call on the UN to Ban 'Killer Robots'"], *Xinhua News*, August 21, 2017.

[99] Zhang Gan, 2018; 赵汀阳 [Zhao Tingyang], "人工智能'革命'的'近忧'和'远虑'—种伦理学和存在论的分析" ["'Short-Term Problems' and 'Long-Term Worries' of the AI 'Revolution'—An Ethical and Existential Analysis"], 哲学动态杂志 [*Philosophical Trends*], June 6, 2018, trans. Jeffrey Ding.

AI magnates wrote to the UN calling for a ban on lethal LAWS appeared in CCP and Chinese government publications, such as 新华 [*Xinhua*] and 科技日报 [*Science and Technology Daily*].[100] While these outlets did not overtly endorse the letter, they gave relatively positive coverage, mentioning Musk's dire warnings and quoting experts who believe that if such systems are ever developed, it will be extremely difficult to ban them. A *Xinhua* article also mentioned that some Chinese people in the tech sector had expressed their support for a ban on killer robots.[101] Zhao Tingyang's article claiming weaponized AI represents a serious threat to humanity was published in a government-affiliated periodical.[102]

Concerns over AI ethics and privacy protection have also been increasing among the population at large, especially since mid-2017. A 2017 poll published in *Beijing Youth Daily* (a CCP newspaper) found that leakage of private data was a top concern among consumers.[103] When China's online retailer Alibaba, clandestinely enrolled people in its Sesame Credit system (which assigns people a trustworthiness score by collecting a variety of data on users' contacts, shopping behavior, finances, and so on) by sneaking the user agreement into an unrelated app in early 2018, the move caused public uproar.[104] In March 2018, a survey by Bytedance, another of China's leading AI firms and creator of the popular news aggregator Jinri Toutiao, found that the two questions that concerned people the most in regard to AI (after which jobs were likely to be automated) were whether it posed a threat and whether AI would be able to have moral or legal awareness.[105]

While the Chinese technologists, academics, and commentators who have begun speculating on AI safety after the release of the Next Generation AI Development Plan are probably not acting under government control, the state has a far greater hand in the overall movement than do Western governments.[106] AI conferences and standards-setting bodies in China are dominated by state organizations, Internet companies, and universities, all with close ties to the government and relatively little input from independent civil organizations.[107] Even in the realm of privacy, where there does seem to have been a genuine wave of public outrage at the more egregious

---

[100] Li Qianying, 2017; 张梦然 [Zhang Mengran], "联合国明年讨论致命性自主武器系统 管控'杀手机器人'" ["Next Year the UN Will Discuss Lethal Autonomous Weapons Systems and Control 'Killer Robots'"]," 科技日报 [*Science and Technology Daily*], December 24, 2016.

[101] Li Qianying, 2017.

[102] Zhao Tingyang, 2018.

[103] Paul Mozur, "Internet Users in China Expect to Be Tracked. Now, They Want Privacy," *New York Times*, January 4, 2018.

[104] Mozur, 2018.

[105] "网民科普阅读大数据报告显示 人工智能成最热话题" ["Big Data Report on the Reading Habits in Popular Science Articles Show AI Is the Hottest Topic"], *Xinhua*, March 28, 2018, trans. Jeffrey Ding.

[106] Jeffrey Ding, Paul Triolo, and Samm Sacks, "Chinese Interests Take a Big Seat at the AI Governance Table," Center for New American Security, June 20, 2018.

[107] Ding, Triolo, and Sacks, 2018.

practices of some Chinese internet companies, the state has stepped in to shape the debate and ensure that it does not interfere with its operations.[108] This is especially true of concerns over military AI, which have tended to hew to the positions Beijing has staked out at the CCW conferences. While some scientists and engineers in the AI field may have moral qualms about developing military AI, they have not organized into anything like the Campaign to Stop Killer Robots in the West, and it is difficult to imagine any large Chinese AI firm refusing a significant military contract under government pressure. In keeping with the government's stated priorities, public discussions of ethics and AI tend to focus more on commercial applications and safety.

One indication of the level of state involvement in discussions over the ethics of AI in China is the fact that it has been government institutions, not civil organizations, that have taken the lead in proposing and drafting safety standards in this field. This is markedly different from U.S. participation in such standards-setting bodies, which are led by the private sector with little government coordination, At the April 2018 meeting of the International Standardization Organization and International Electrotechnical Commission (jointly responsible for about 85 percent of international product standards), Joint Technical Committee 1, Subcommittee 42 on AI in Beijing, the Chinese delegation was led by the China Electronic Standards Institute (CESI), a standards-setting body directly under the Chinese Ministry of Industry and Information Technology. A CESI-led coalition of major Chinese tech companies and universities drafted a white paper for the conference, setting forth proposed standards for AI products and services.[109] CESI noted four key ethical principles that must be foundational for AI systems, including the centrality of human interest in system design, the creation of clear liability when systems fail, transparency in the operation of these systems, and the need to balance the responsibility for transparency with the needs of companies to keep trade secrets. The need to maintain privacy was also emphasized. The paper listed "military robots" as an AI product, and so it is likely that whatever standards it develops could have some applicability in the military realm.[110] As of August 2019, Subcommittee 42 is preparing an "overview of ethical and societal concerns" as well as an overview of risk management. An official document on bias in AI decisions has been proposed, and a draft overview of trustworthiness of AI systems is being voted on in committee.[111] CESI and other state-led organizations will likely continue to develop domestic AI safety, liability, and ethical standards and seek to internationalize these standards. Once again, Beijing

---

[108] See section on privacy below.

[109] Ding, Triolo, and Sacks, 2018.

[110] 中国电子技术标准化研究院 [China Electronic Standardization Institute], 人工智能标准化白皮书 [AI Standardization White Paper], 2018. Jeffrey Ding and Paul Triolo did a partial translation of this piece, available at https://www.newamerica.org/cybersecurity-initiative/digichina/blog/translation-excerpts-chinas-white-paper-artificial-intelligence-standardization/.

[111] "ISO/IEC JTC 1/SC 42: Artificial Intelligence," International Organization for Standardization, n.d.

has publicly committed to developing ethical standards for AI, but these will be government standards, with minimal oversight by any domestic organization.

Privacy—Strong Standards for Business, Not for the Government

The Chinese public may be more involved in the debate over privacy rights than in any other facet of AI-related ethics, yet the role of government-led organizations in the discussion still looms large. For example, the consumer protection organization that sued Baidu in early 2018 for illegally collecting personal information was government-backed.[112] The National Information Security Standardization Technical Committee released Personal Information Security Technology Standards in 2018. These standards provide extensive protections to Chinese consumers, including guarantees that any organization using their data may use them only for purposes related to the service for which they were collected and must delete or anonymize the data once that purpose has been achieved.[113] The standards also reiterate earlier Chinese legislation preventing the transfer of data on Chinese consumers to entities outside of China.[114] These protections are in many ways comparable to Europe's stringent General Data Protection Regulation.[115] But while the standards may protect users from private companies, they contain wide loopholes for involuntary data collection on any projects relating to national or public security.[116] Thus, while Chinese private enterprise could increasingly be on a more equal footing with their U.S. and European counterparts in terms of the legal restrictions they have on data use (they may actually be at a slight disadvantage given the United States' lack of similar privacy legislation), the PLA is likely to continue to enjoy a more unfettered ability to use its citizens' data than DoD.

Conclusion

The PLA is aggressively developing robotic and software AI applications. There are no known Chinese weapon systems currently fielded or under development that are fully autonomous, but some are already highly autonomous and could likely operate without a human in the loop if their software were modified. Were the PLA to go that route, it would face the same challenges that the United States does in determining responsibility and accountability for the behavior of LAWS. In fact, these challenges may be even greater for China, given the PLA's

---

[112] Samm Sacks, "New China Data Privacy Standard Looks More Far Reaching Than the GDPR," Center for Strategic and International Studies, January 29, 2018.

[113] 全国信息安全标准化技术委员会 [National Information Security Standardization Technical Committee], 信息安全技术 个人信息安全规范 [Personal Information Security Information Security Technology Standards], January 24, 2018.

[114] National Information Security Standardization Technical Committee, 2018.

[115] Sacks, 2018.

[116] National Information Security Standardization Technical Committee, 2018.

traditional reluctance to delegate authority. The PLA is currently conducting research in ML target-identification technologies, which could make their weaponry even more independent of operators. Beijing is also working on a variety of AI and ML-based systems to autonomously integrate data from a wide variety of sensors to identify hidden targets, provide a common operating picture to commanders, and enable rapid decisionmaking. In the future, the PLA hopes that this system will act as a "digital staff officer" to provide information, recommendations, and planning suggestions to commanders.

Although Beijing has officially called for a ban on LAWS, the way it proposes to define that class of weapons is so narrow that even if such a ban was approved, it would likely do little to affect any system currently under development in either the United States or China. The Chinese public has become increasingly engaged in the debate over the morality of military AI, but that engagement is dominated by state and state-affiliated institutions. Whatever qualms Chinese academics and technology magnates have, they are unlikely to restrain Beijing's actions.

Nevertheless, the United States has little to lose in engaging China in negotiations on an international agreement to restrict the development or employment of LAWS. Beijing's proposed ban does not appear to be a disguised attempt to constrain the United States' AI development, because U.S. standards are already stricter than those in the proposed ban. China is more likely to moderate its behavior if there is an international treaty in place requiring some level of human control. The United States might thus benefit from entering into negotiations and trying to ensure that the definition of LAWS finally adopted by the UN is as close as possible to DoD ethical guidelines on autonomous weapons.

6. Military Artificial Intelligence in Russia

Russia is devoting considerable energy to leveling the playing field with the United States in military AI. As mentioned in Chapter 2, President Vladimir Putin has stressed the gravity of the situation, warning that military AI has the potential to disrupt strategic stability if Russia falls behind: "Artificial intelligence is the future, not only for Russia, but for all humankind. It comes with colossal opportunities, but also threats that are difficult to predict. Whoever becomes the leader in this sphere will become the ruler of the world."[1]

To close the gap and remain competitive, the Russian defense ministry has responded with a flurry of activity to improve its military AI. This chapter will outline some of Russia's key recent and projected developments in this area, and it will consider how those developments might be influenced by past brushes with autonomy. It will also explore the cultural, structural, and demographic factors that could help or impede progress and describe the direction of Russian policy on the appropriate use of military AI.

Current Capabilities and Future Projections

Like the United States and China, Russia is actively exploring how AI can improve operational effectiveness and increase efficiency across a range of military applications. Still in the nascent stages, AI R&D in Russia remains broad. For now, decisionmakers are withholding judgment and not ruling out any options for future development before they thoroughly understand the benefits that AI might provide.

This strategy parallels Russia's flexible approach to warfare. By not preferring one form of warfare over another, Russia attempts to tailor solutions specifically to the given circumstances. As David Shlapak has said, "If 'little green men' will do the job, then 'little green men' will be employed; if big green tanks are needed instead or in addition, bring on the big green tanks."[2] Russia's ability to project power has continued to diversify in recent years as it has managed to expand its nuclear deterrent, maintain a looming conventional threat in the Baltics, and develop new instruments of influence in what has come to be called the "gray zone."[3] There is evidence that AI is permeating into all of these spaces and more.

---

[1] "'Whoever Leads in AI Will Rule the World': Putin to Russian Children on Knowledge Day," *RT International*, September 1, 2017.

[2] David A. Shlapak, *The Russian Challenge*, Santa Monica, Calif.: RAND Corporation, PE-250-A, 2018.

[3] According to Hal Brands, "Gray zone conflict is best understood as activity that is coercive and aggressive in nature, but that is deliberately designed to remain below the threshold of conventional military conflict and open interstate war." See Hal Brands, "Paradoxes of the Gray Zone," Foreign Policy Research Institute, February 5, 2016.

Unmanned Vehicles

Russia is dedicating significant resources to developing unmanned systems, as evidenced by the host of new UAVs, UGVs, and unmanned undersea vehicles (UUVs) in various phases of R&D and testing. Most of these "combat robots," or "intelligent robotic complexes," as they are called in Russia, are still remote-controlled, but Russian military observers note that the level of autonomy can be scaled as the software improves.[4] The arms manufacturer Kalashnikov Concern seems to be at the cutting edge, claiming it has developed a "combat module" that is based on neural networks and is capable of autonomous target identification and decisionmaking.[5] There are other UGVs, such as the *Nerekhta*, that can allegedly navigate to predetermined targets without a remote operator.[6] Then there is the formidable *Uran-9*, an unmanned tank housing a 30-mm 2A72 automatic cannon, a 7.62-mm machine gun, and anti-tank guided missiles, which can be operated by remote control or can function autonomously.[7] Despite reports claiming that it underperformed during testing in Syria, the *Uran-9* remains a closely watched UGV because there is no direct analog to it in the U.S. military.[8]

The head of robotics at the Skolkovo Innovation Center's Information Technology "Cluster" says that, in addition to offensive combat operations, Russia may consider unmanned vehicles for reconnaissance and surveillance, patrol and fire support, sentry and site security, ammunition delivery, casualty evacuation, installing minefields and demining, cover and concealment, and even providing mobile audio propaganda.[9]

So far, most of these combat robots have not been fully integrated into the armed forces, but there are reasonable arguments for Russia to continue taking steps to do so. Three of the Russian SMEs whom we interviewed mentioned that deploying more robotics would help mitigate manpower and budget constraints in the Russian military. As one of them said, "[The Russians] think the development of more unmanned systems will help them level the playing field. Demographically, looking out to 2035, it's not a catastrophe. But the population won't be growing very much, so they're thinking about that manpower constraint creatively."[10] As for costs, Russia has a smaller defense budget than the United States or China; having fewer soldiers in the ranks would go a long way to reducing the long-term financial burdens of training, health care, pensions, and other personnel costs. Moreover, having a greater proportion of robotic

---

[4] Vadim Kozyulin and Albert Efimov, "Новый Бонд—Машина С Лицензией На Убийство" ["The New Bond: A Machine with License to Kill"], Индекс Безопасности [*Security Index*], Vol. 22, No. 1, pp. 34–35.

[5] "Kalashnikov Gunmaker Develops Combat Module Based on Artificial Intelligence," *TASS*, July 5, 2017.

[6] Mikhail Nekrasov, "Russian Military Robots in Action," *Russia Beyond*, January 11, 2017.

[7] Kozyulin and Efimov, 2018.

[8] "Combat Tests in Syria Brought to Light Deficiencies of Russian Unmanned Mini-Tank," *Defense Blog*, June 18, 2018.

[9] Kozyulin and Efimov, 2018.

[10] Interview with Dara Massicot, Santa Monica, Calif., May 7, 2018.

systems would concentrate more power in the hands of fewer people. This would help to alleviate concerns about disloyalty or incompetence in the lower ranks.

With these benefits in mind, Russian leaders have developed concrete goals for the military's robotization. Defense Minister Sergei Shoigu hopes that serial production of combat robots will begin in 2018.[11] And Russia's Military Industrial Committee has "set a target of making 30 percent of Russia's military equipment robotic by 2025."[12] Such goals are ambitious, considering some of the roadblocks Russia has run into in developing unmanned systems thus far. Most notably, Russia has still not managed to field its first medium-altitude long-endurance unmanned combat aerial vehicle (UCAV).[13] In comparison, the MQ-1 Predator has been used for combat operations in the U.S. military since 2001.[14] That puts Russia 17 years behind. Of course, it bears repeating that these are remotely piloted aircraft that would not qualify as AI systems by any reasonable definition of the term. Still, the comparison carries some weight since this kind of UCAV is probably a precursor to other more sophisticated technologies.

Swarm technology in Russia is also a long way off compared with the United States and China. Although the Russian military has shown interest in having its robots coordinate through networks, the subject is usually discussed in more general and aspirational terms. The former commander of the Russian Aerospace Forces, General Viktor Bondarev, mused, "Flying robots will be able to act in formation, not separately. Perhaps the operator will sit on the ground and monitor the whole unmanned squadron with a computer."[15] For autonomous underwater vehicles (AUVs), there is talk of developing mini-torpedoes that slowly move through the water in groups. Such weapons would be silent and inconspicuous, traveling at just 2–3 miles per hour while using AI to alter their movement patterns in a way that resembles a school of fish or "combat turtles." This concept is still in its infancy and at least ten years from practical implementation, but one of Russia's leading missile designers is optimistic that the idea will progress.[16]

Even if this technology does not pan out, Russia has a completely different AUV design that has undergone successful testing and is potentially more troubling than any of the autonomous systems mentioned thus far. The *Status-6* AUV is one of Russia's newest nuclear delivery systems. According to President Putin, it can dive to "great depths," travel at a speed "multiple

---

[11] "Combat Robots for Russian Troops to Go into Serial Production This Year—Defense Minister," *TASS*, March 15, 2018.

[12] Tom Simonite, "For Superpowers, Artificial Intelligence Fuels New Global Arms Race," *Wired*, September 8, 2017.

[13] Samuel Bendett, "Red Robots Rising: Behind the Rapid Development of Russian Unmanned Military Systems," *Strategy Bridge*, December 12, 2017.

[14] James Thompson, "Sunsetting the MQ-1 Predator: A History of Innovation," *U.S. Air Force*, February 20, 2018.

[15] "Россия Полагается на Боевых Роботов" ["Russia Relies on Combat Robots"], *Калашников Медиа* [*Kalashnikov Media*], November 20, 2017.

[16] "Шамиль Алиев: торпеду 'Шквал' надо опустить на глубину" ["Shamil Aliyev: The Torpedo 'Shkval' Should Be Lowered to the Deep"], *РИА Новости* [*RIA Novosti*], September 11, 2017.

times higher than the speed of submarines, cutting-edge torpedoes, and all kinds of surface vessels," and carry either a conventional or nuclear warhead.[17] The AUV is meant to idle within range of its target—most likely a coastal city or aircraft carrier—and wait for a remote signal to attack.[18] This method of delivering nuclear weapons is particularly worrisome because the AUV cannot reliably be recalled post-launch, given how difficult undersea communications are.[19]

Defensive Systems

Defensive systems with varying degrees of autonomy have been in service in Russia for decades, starting in 1978, when the Soviet Union became the first country to build an active protection system (APS) for armored vehicles. *Drozd* was a hard-kill APS meant to stop incoming anti-tank guided missiles and grenades. Despite its tendency to cause collateral damage, *Drozd* was installed on over 250 T-55A tanks. In 1993, Russia developed a new hard-kill APS called *Arena*. Like all APSs, *Arena* had to operate in a fully autonomous mode so that it could react quickly enough to incoming threats.[20] Now, Russia's newest APS, *Afghanit*, is being added to the T-14 Armata tank.[21] It will provide 360-degree coverage and be able to simultaneously detect and track up to 40 ground targets and 25 aerial targets.[22]

Russia's S-400 Triumf surface-to-air missile (SAM) system functions in a similar role as an APS, performing both autonomous detection and targeting of incoming airborne threats. Like an APS, the system also has to be autonomous for it to function most quickly and effectively. The S-400 can simultaneously engage up to 36 targets at a distance of up to 250 kilometers.[23]

In addition, Russia has made strides to improve the integration of its air defense systems by using AI to reduce the cognitive burden on soldiers. Currently, Russia's SAMs and radars work independently, forcing soldiers at command posts to collate multiple information streams while they assess targets. But in the spring of 2018, Russia began testing a new automated control system (ACS) that unifies the S-300 and S-400 SAM batteries, the Pantsir-S antiaircraft weapons, and modern radar systems under a single domain. Using "elements of artificial intelligence," the ACS will offer commanders a complete picture of the air situation with options

---

[17] Vladimir Putin, "Presidential Address to the Federal Assembly," March 1, 2018.

[18] "Океанская многоцелевая система 'Статус-6' (Kanyon)" ["Ocean Multi-Purpose System 'Status-6' (Kanyon)"], *Новости ВПК* [*Novosti VPK*], November 27, 2016.

[19] "Just How Much of a Threat Is Russia's Status-6 Nuclear Torpedo?" *National Interest*, January 16, 2018.

[20] Tom J. Meyer, "Active Protective Systems: Impregnable Armor or Simply Enhanced Survivability?" *ARMOR*, May–June 1998.

[21] It is worth noting that the T-14 Armata has an unmanned turret as well.

[22] Vincent Boulanin and Maaike Verbruggen, "Mapping the Development of Autonomy in Unmanned Systems," Stockholm International Peace Research Institute, November 2017, p. 43.

[23] Boulanin and Verbruggen, 2017, p. 37.

for attacking incoming threats in a way that promises to maximize effectiveness, efficiently distribute resources, and accelerate the decisionmaking process.[24]

Planning and Decisionmaking Tools

Beyond robotic platforms, we are also starting to see elements of AI appear in Russia's C2 infrastructure, particularly in planning and decisionmaking support platforms. This should not be surprising, given Russia's historical inclination to use mathematical models and other quantitative tools to inform military decisions. Dating back to the Soviet era, computer programs have had a surprising amount of influence on strategy.

One particularly illustrative example is the Raketno-Yadernoye Napadenie (RYAN) computer model which was developed by the Committee for State Security (Komitet Gosudarstvennoy Bezopasnosti [KGB]) in the early 1980s to calculate the overall strategic balance between the Soviet Union and the United States.[25] The model used a steady stream of intelligence data to determine the correlation of forces between the superpowers. The results were grim and stoked a growing fear in the upper echelons of Soviet leadership that the United States was on the brink of achieving decisive military superiority. Soviet analysts concluded that if the power imbalance became too great, they would be forced to initiate a preemptive first strike before the Americans did the same. The threshold for this decision hinged on a critical value derived from RYAN: According to intelligence archives, if the Soviet's strength dropped below 40 percent relative to the United States, the KGB and military leaders would advise the Kremlin to launch a preemptive nuclear attack.[26]

Luckily that was never necessary. Still, the Politburo's dependence on RYAN was remarkable. While RYAN is now a 35-year-old Cold War relic, one expert we interviewed said that even today, "Russia wants to make war a math problem."[27] There might be some truth to that. Building on progress made in the 1980s, the Russian military has steadily developed a composite system of mathematical models to support operational and strategic planning. The system's main functions are to determine optimal force distributions, track troop movement, and ultimately forecast the course and outcome of operations. Already, the system has performed so

---

[24] Alexander Kruglov, Alexey Ramm, and Evgeny Dmitriev, "Средства ПВО объединят искусственным интеллектом" ["Air Defense Weapons Will Be Combined with Artificial Intelligence"], Известиыа [*Izvestiya*], May 2, 2018.

[25] "Raketno-Yadernoye Napadenie" is Russian for "Nuclear-Missile Attack." For a detailed discussion on the KGB's development and use of RYAN to calculate the correlation of forces between the superpowers in efforts to anticipate a U.S. attack on the Soviet Union, see Benjamin B. Fischer, "A Cold War Conundrum: The 1983 Soviet War Scare," *Library*, Central Intelligence Agency, July 7, 2008. Some sources add "Vnezapnoe" [Surprise] to the title, making the acronym "VRYAN," subsequent to the publication of a book by a former KGB officer that uses that name for the model. See Yuri B. Shvets, *Washington Station: My Life as a KGB Spy in America*, New York: Simon & Schuster, 1994, p. 74.

[26] President's Foreign Intelligence Advisory Board, "The Soviet 'War Scare,'" February 15, 1990, p. 45.

[27] Interview, name withheld upon request.

successfully in training events that some officers claim it could operate relatively autonomously in many military command centers.[28]

The newly built National Defense Management Center (NTsUO) might be a current location or future destination for this system. The NTsUO was activated in 2014 in an effort to create a unified information space for the Russian national security apparatus. While exact details regarding activity at the NTsUO are not disclosed, the Russian defense ministry has said that it will collect, collate, and analyze information on the "military-political situation" in the world, as well as on strategic directions and on the sociopolitical situation during peacetime and wartime. Like the decision support system described above, it will monitor the readiness state of the armed forces and the strategic grouping of troops.[29] There is no direct link between AI and the activity at the NTsUO, but it is notable that about half of the Russian state's investments in the sphere of AI are dedicated to data analysis projects (33 percent) and decision support systems (16.5 percent).[30]

Gray-Zone Activities

Before AI appears in full-scale conflicts, it will likely be employed in ambiguous situations that may not constitute either war or peace. In recent years, Russia has attempted to exploit these uncertain conditions by applying pressure on other states without formally entering wars. Broadly defined, these "measures short of war" or "gray-zone activities" are coercive techniques that may or may not involve combat but typically blur across diplomatic, informational, economic, and military lines.[31]

Increasingly, they are playing out in the digital domain. Cyberwarfare, electronic warfare (EW), influence operations, propaganda campaigns, and disinformation are prime examples of instruments that fit the Russian modus operandi and are ripe to be integrated with AI. Kremlin-supported groups have already engaged in variations of these activities by spreading false stories on social media, meddling in elections, and infiltrating the control systems of critical infrastructure.[32] Many of the experts we interviewed warned that Russia is almost certainly

---

[28] S. V. Yashin, V. N. Denisov, O. V. Sayapin, and L. V. Makarciev, "Апробация Модели Применения Межвидовой Группировки Войск (Сил) На Стратегическом Командно-Штабном Учении 'Кавказ-2016'" ["Praise for the Model Taking the Inter-Service Correlation of Forces (Strength) in the Strategic Command Post Exercise 'Kavkaz-2016'"], *Военнаыа Мисл* [*Voennaya Misl*], No. 2, February 2018, pp. 28–32.

[29] Mikhail Evgenyevich Mizintsev, "Национальный Центр Управления Обороной Российской Федерации" ["National Center for Defense Management of the Russian Federation"], Ministry of Defense of the Russian Federation, n.d.

[30] "Искусственный Интеллект Научился Привлекать Господдержку" ["Artificial Intelligence Has Learned to Attract State Support"], *Коммерсант* [*Kommersant*], April 4, 2017.

[31] Christopher S. Chivvis, "Understanding Russian 'Hybrid Warfare' and What Can Be Done About It: Addendum," Santa Monica, Calif.: RAND Corporation, CT-468/1, 2017.

[32] Alina Polyakova and Spencer P. Boyer, "The Future of Political Warfare, Russia, the West, and the Coming Age of Global Competition," *Foreign Policy at Brookings*, March 2018.

attempting to leverage AI to increase the quality and scale of these operations. Kiril Avramov from the University of Texas proposed that, instead of paying people at a "troll factory" in St. Petersburg or Macedonia to spread online propaganda, Russia could make the process autonomous with well-trained bot networks.[33]

Additionally, AI software has now made it possible to create ultrarealistic depictions of events that never happened. The experts we interviewed stressed the massive risk posed if Russia began doctoring images and videos for political gain.[34] Most likely, willingness to use this particular technique is an area where Russia and the United States diverge ethically. As we shall discuss in Chapter 7, our survey results showed that more than three-quarters of U.S. respondents found it ethically impermissible for the U.S. military to create fake audio and video footage of a foreign leader in a compromising situation. Of course, we did not poll the Russian public and cannot know its stance on the issue. But compared with the U.S. military, which has committees of elected officials responsible for providing military oversight, the Russian armed forces are not nearly as bound to public scrutiny.

For what it is worth, the experts we interviewed asserted that Russia would have absolutely no compunction about exploiting this technology to push false narratives. Such strong suspicions seem warranted, given Russia's history of engaging in political warfare. The Soviet Union sought to undermine democracy for decades through "active measures" over print and radio. Now they manifest on social media sites. As Avramov observes, "We have old wine in new bottles."[35] Reinforcing this point, the European Union's disinformation watchdog, East StratCom Task Force, recently announced that Kremlin-backed trolls have already started experimenting with rudimentary versions of these "deepfakes."[36]

Moscow has also used the gray zone as a de facto testing ground for new technologies, especially those that may be ethically contentious, because of how hard it is to trace actions directly back to the Russian state. A notable example of this is cyberattacks carried out by proxy groups using commercial or personal internet protocol addresses, which makes attribution notoriously difficult. Trying to attain incontrovertible cyber forensic evidence is fraught with technical and legal hurdles, and the addition of third-party hackers muddles the situation even more.[37] Consequently, the United States and others targeted in such attacks should not set an unattainable standard of conclusive proof before confronting Russia or other perpetrators over

---

[33] Kiril Avramov, interview with authors via telephone, May 14, 2018.

[34] As we mention below, these activities have come to be called "deepfakes."

[35] Avramov interview, 2018.

[36] The term "deepfake" is a portmanteau of "deep learning" and "fake." See the following articles for more information: Scott Edwards and Steven Livingston, "Fake News Is About to Get a Lot Worse. That Will Make It Easier to Violate Human Rights—And Get Away With It," *Washington Post*, April 3, 2018; Nick Harding, "'Deepfake' Videos Produced by Russian-Linked Trolls Are the Latest Weapon in Fake News War, Official Monitors Warn," *The Telegraph*, May 26, 2018.

[37] Pierluigi Paganini, "Cyber Warfare: From Attribution to Deterrence," Infosec Institute, October 3, 2016.

their misbehavior. Yet, even if Russia is the obvious culprit, there are not well-established retaliation measures in place to deter future action.

Given the current uncertainty about what states can and cannot get away with, Russia is able to press ahead. At a military-scientific conference in March 2018 called "Artificial Intelligence: Problems and Solutions," Deputy Defense Minister Yuri Borisov stated in his opening remarks that the development of AI technologies will provide Russia with the necessary means to counteract opposition in the information space and win cyberwars. According to Borisov, all battles are first played out in the information space rather than on the physical battlefield. He who can control the information space and properly manipulate the opposition becomes the winner. Borisov referred directly to cyberwarfare in his speech, but his message could apply to EW, influence operations, disinformation, and more.[38]

Russia has recently developed a new C2 platform for EW at the brigade level, called "Bylina." It is a fully autonomous system that analyzes combat situations, identifies targets, chooses how to disable them, and ultimately issues orders to EW forces in the field. Human officers in the brigade will provide oversight by monitoring the system's performance from the battlefield's periphery. Procurement of Bylina will begin in 2018, with 2025 as the target date for its arrival in all brigades.[39]

Russia has made Bylina, and EW more generally, a priority because of its potential to negate U.S. information superiority on the battlefield. Colonel Yuri Gubskov, who oversees the interservice center for EW training, has indicated that Bylina is just the tip of the iceberg; a host of robotic systems could have similar AI algorithms in the near future. He says, "The development of information technologies, which can be used for EW, allows us to talk about the possibility of soon creating robotic complexes with elements of AI. New systems will be able to effectively solve problems in a complex radio-electronic environment without human participation."[40]

Translating the Vision to Reality

So far, we have discussed what Russia hopes to gain from integrating AI into the military. For some systems, such as Bylina, sufficient documentation is publicly available to paint a reasonably accurate picture of what they entail, but the sources for other systems are less reliable, and claims could be exaggerated. Because of these uncertainties, it would be unwise to

---

[38] Yuri Avdeev, "На Пути К Искусственному Интеллекту" ["Toward Artificial Intelligence"], *Красная Звезда* [*Red Star*] March 16, 2018.

[39] Alexey Ramm, Dmitry Litovkin, and Evgeny Andreev, "На Пути К Искусственному Интеллекту" ["The Forces of Electronic Warfare Will Bring Artificial Intelligence"], *Известиыа* [*Izvestiya*], April 4, 2017.

[40] Alexey Ramm, "Новые Системы Смогут Эффективно Решать Задачи Без Участия Человека" ["New Systems Can Effectively Solve Problems Without Human Intervention"], *Известиыа* [*Izvestiya*], April 27, 2018.

speculate. Instead, it is preferable to step back and frame the discussion of Russian AI—and Russian military innovation more generally—in terms of the structural and cultural factors that tether them to reality.

In addition to directly comparing military systems of different countries, understanding the underlying mechanisms that produce them is beneficial. In particular, access to funding, resources, infrastructure, and human capital are key factors influencing whether Russia has the ability to excel in AI development.

Modest but Consistent Funding

Overall, the Russian Ministry of Defense (MoD) has had fairly consistent funding in recent years and seems financially stable for the near future. Investment spending, which constitutes over half of the annual MoD budget, is probably safe.[41] For the whole country, Systems, Applications, and Products in Data Processing (SAP), a multinational software and information services company based in Walldorf, Germany, reported that there were 1,386 scientific projects in AI from 2007 to 2017. Of these, almost 90 percent were state-funded, amounting to 22.9 billion rubles, or roughly $363.5 million, spent by the government over a ten-year period.[42] That averages to a $36-million annual state investment in AI, which is high compared with a U.S. source that estimates the market size was less than $12 million in 2017.[43] Either way, the $12–$36 million range should not be considered very high, given that the U.S. government allocates about $200 million into AI research annually.[44]

That said, statistics on Russian defense spending are not entirely reliable, given that much of the MoD R&D budget is classified, making it difficult to reliably track the exact flow of money into AI technology. For example, funding for the development of the *Status-6* intercontinental AUV would not have been captured by this analysis.

Streamlining and Standardizing Efforts

Until 2014, Russia lacked a true strategy for military robotics and AI more generally. Designs were disparate and uncoordinated, and state policies, standards, and testing procedures were not in place to smoothly usher programs from conceptual prototypes into established

---

[41] Brooks Tigner, "Russia's 2017 Defense Spending Cut Is Not What It Seems," Atlantic Council, May 9, 2018.

[42] "Искусственный Интеллект Научился Привлекать Господдержку" ["Artificial Intelligence Has Learned to Attract State Support"], 2017. Estimation calculated based on March 31, 2017, ruble-to-dollar exchange rate. Also see "Где Найти Интеллект для Бизнеса" ["Where to Find Intelligence for Business"], SAP Planet, 2017.

[43] Jill Dougherty and Molly Jay, "Russia Tries to Get Smart About Artificial Intelligence," *Wilson Quarterly*, Spring 2018.

[44] "Искусственный Интеллект Научился Привлекать Господдержку" ["Artificial Intelligence Has Learned to Attract State Support"], 2017.

systems. CNA's Samuel Bendett summarized how a directed campaign helped rapidly address these key deficiencies:

> In 2014, the Russian Ministry of Defense developed and approved a comprehensive target program called "Creation of Prospective Military Robotics through 2025." The Ministry also formed a commission for the development of robotics, headed by the Defense Minister Sergei Shoigu. To formulate battlefield needs for the next 10–20 years and to justify developments of military robotics, Russia launched an annual conference in 2016 called "Robotization of the Armed Forces of the Russian Federation." The goal of this annual event was the development of "unified interdepartmental approaches for the creation and development of military and special-purpose robotic complexes (RTCs)." Russia also launched its own version of the Defense Advanced Research Projects Agency (DARPA) called Foundation for Advanced Studies, tasked with working on various unmanned and robotics projects for the military.[45]

Additionally, in 2016 the state implemented new standards for testing, requirements formation, and technical documentation of military robotics.[46] Likewise, in 2018 the Foundation for Advanced Studies "prepared proposals for the standardization of artificial intelligence for the [MoD]."[47]

These actions will undeniably help facilitate the management of military robotics, but it is revealing that this inflection point occurred only recently. Russian leaders are eager to see improvements on this front continue, as evidenced by President Putin's statement in May 2018 calling for additional streamlining: "As soon as possible, we need to develop a progressive legal framework and eliminate all barriers for the development and wide use of robotic equipment, artificial intelligence, unmanned vehicles, e-commerce and Big Data processing technology. And this legal framework must be continuously reviewed and be based on a flexible approach to each area and technology."[48]

Innovation in the Private and Public Sector

MoD is well aware that success will depend largely on the state's ability to harness talent from the private sector. On behalf of Defense Minister Shoigu, Deputy Defense Minister Borisov called on military and civilian scientists to "unite [their] efforts in the research, development, and

---

[45] Bendett, 2017.

[46] Sergey Popov and Oleg Falichev, "Робот Стреляет Первым" ["The Robot Shoots First"], *Новости ВПК* [*Novosti VPK*], February 23, 2016.

[47] Sergei Barbuk, "ФПИ Предложил Минобороны Стандарты Для Искусственного Интеллект" ["FPI Offered the Ministry of Defense Standards for Artificial Intelligence"], *Новости ВПК* [*Novosti VPK*], March 20, 2018.

[48] Putin, 2018.

implementation of artificial intelligence technologies."[49] To facilitate that unity directly, Borisov introduced plans to construct an entire city dedicated to military innovation, in which AI will be a specific focus area. By 2020, this "technopolis" will house 80 scientific and industrial enterprises and have as many as 2,100 specialized jobs in research, experimentation, and testing.[50] On a separate front, the Russian MoD is now holding AI design competitions that are open to the public in the style of DARPA Grand Challenges.[51]

Despite small similarities, the experts we interviewed pointed out that the relationship between the public and private sectors in Russia is far different than in the United States. Russia's system is one of almost complete state control of the defense industrial base, primarily through Rostec, a defense conglomerate comprised of various holding companies. Since the defense industrial base is consolidated under government control (and President Putin has direct oversight as chairman of the Defense Industrial Commission), it is very responsive to demand.[52] When the military identifies a need, it can rapidly shift its focus of research in a new direction.

The downside is that while this arrangement arguably adds more flexibility to the R&D process, it creates an environment starved of competition. Russia's defense industrial base is more similar to an oligopoly than a perfectly competitive market, and without competition driving innovation forward, there is less incentive to develop the best products. Two interviewees suggested that Russia's trade-off between responsiveness and competitiveness could be a limiting factor for the development of complex AI technologies.

Human Capital

The nature of the civil-military relationship is meaningless if not buttressed by an educated and productive workforce. Taking a deep dive into Russian demographics, the Jamestown Foundation found that Russia is pretty well educated. It performs on par with most of Western Europe in mean years of schooling, and almost 60 percent of Russians 25 or older have some form of higher education.[53]

Yet Russia has paradoxically high levels of mortality and gross domestic product per capita, on par with some less-developed countries. Jamestown also found that Russia's knowledge production, estimated by number of patents earned, is dismal: "The entire Russian Federation did

---

[49] Sergei Shoigu, "Шойгу Призвал Военных И Гражданских Ученых Совместно Разрабатывать Роботов И Беспилотники" ["Shoigu Urged Military and Civilian Scientists to Jointly Develop Robots and Drones"], *TACC* [*TASS*], March 14, 2018.

[50] Avdeev, 2018.

[51] Samuel Bendett, "The Russian Military Wants Students to Design Its New Underwater Drone," *War Is Boring*, March 7, 2018.

[52] "Putin Takes Greater Control of Russia's Defense Sector," *Stratfor*, September 8, 2014.

[53] Nicholas Eberstadt, "Demography and Human Resources: Unforgiving Constraints for a Russia in Decline," *Jamestown Foundation*, September 13, 2016. Mean years of schooling measures the average number of years of education received by people 25 years or older.

not earn as many patents as the U.S. state of Alabama between 2001 and 2015—and Alabama's population is scarcely more than a thirtieth of Russia's."[54]

Russia's knowledge production in the field of AI is similarly unremarkable. Whereas China and the United States rank first and second, respectively, in the world for the number of AI publications and citations from 1996 to 2017, Russia resides much further down the list at thirty-third and forty-second, respectively.[55] Despite an uptick in AI publications since 2012, Russia's citation rate has steadily dropped. This indicates a negative correlation between the volume and quality of papers.[56]

Russia's apparent inability to translate education into knowledge production could partially be explained by the steady migration of educated people out of Russia. The Russian Presidential Academy of National Economy and Public Administration estimates that roughly 100,000 Russians emigrate to developed countries every year and that around 40 percent have a higher education.[57] The same study found that exactly half of all doctoral students hoped to leave the country to find a "good job."[58]

However, the structural, cultural, and demographic problems facing Russia should not be overstated. Yes, there are factors that indicate it will be an uphill battle for Russia to come close to reaching parity with the United States and China in AI. However, it is incorrect to claim that all AI is expensive to develop or demands a large coordinated effort from the state. As one interviewee suggested, "a couple of guys in their basement" could single-handedly doctor videos or train an army of internet bots.[59]

Thus, it is certainly possible that Russia could take a leaner approach to AI development in the long term. Low-cost gray-zone applications would remain available, even if combat robots, control systems, and complex decisionmaking tools become resource-prohibitive.

But that would be getting ahead of ourselves. For now, nothing is off the table.

Ethical Considerations

How Russia's leadership comes down on the ethical questions presented by military AI remains unclear. At the very least, President Putin recognizes that the development of AI presents "threats that are difficult to predict."[60] Yet, Russian leaders have not publicly addressed humanitarian concerns raised by organizations such as the Campaign to Stop Killer Robots and

---

[54] Eberstadt, 2016.

[55] "Scimago Journal and Country Rank," *Scimago*, n.d.

[56] "Startup Investment & Innovation in Emerging Europe," *East-West Digital News*, Version 1, February 2018.

[57] "Russia's Brain Drain on the Rise over Economic Woes—Report," *Moscow Times*, January 24, 2018.

[58] "Half of Russian PhD Students Want to Move Abroad," *Moscow Times*, April 4, 2018.

[59] Rand Waltzman, interview with the authors, Santa Monica, Calif., March 8, 2018.

[60] "'Whoever Leads in AI Will Rule the World,'" *RT International*, September 1, 2017.

the Future of Life Institute. It is likely that Russia is more concerned with the strategic and operational risks presented by AI than with ethical considerations.

Ethics Is a Low Priority

Samuel Bendett suggests that it would be premature for Russia to talk about ethics when autonomy in the Russian military is still very limited. He says that at present, "Russia is not looking at [AI] as an ethics issue."[61] The white paper that Russia submitted at the 2018 UN CCW GGE meeting matches this tone of ethical indifference, inasmuch as it urges the GGE not to define LAWS in a way that would undermine ongoing research in AI or lead to a debate about "good" versus "bad" weapons.[62]

That is not to say Russia is devoid of humanitarian concerns. In the same paper, Russia notes the potential for AI to improve the precision of its weapons and reduce the risk of civilian collateral damage.[63] However, it is not clear that ethics are the primary motivating factor. According to Bendett, the main driver for reducing humanitarian harms is to improve perceptions of Russia abroad.[64] The analysts we interviewed shared the sentiment that ethics is not a priority in Russia's military plans for AI. Multiple interviewees expressed the opinion that catching up with the United States in AI dwarfs all of Russia's other concerns.

The experts we interviewed noted that they have not seen widespread discussions of ethical issues from Russian scientists or business leaders either. "The worries we have [in the United States], espoused by voices like Elon Musk and Google, are not present in Russia. Or, at the very least, this is something that doesn't publicly surface in Russian discourse," said Avramov.[65] This point was reinforced in a military journal article by Igor Popov, a retired Russian colonel. After grappling with the pros and cons of granting an autonomous weapon the right to kill, Popov laments, "The problems of future wars are not discussed at scientific conferences or the military media. . . . The public does not receive any information on these issues from the Ministry of Defense. . . . I really want to believe that our generals know the answers to these questions, but do not consider it necessary to inform a wide audience."[66]

---

[61] Bendett, 2018.

[62] "Russia's Approaches to the Elaboration of a Working Definition and Basic Functions of Lethal Autonomous Weapons Systems in the Context of the Purposes and Objectives of the Convention," Group of Governmental Experts of the High Contracting Parties to the Convention on Prohibitions or Restrictions on the Use of Certain Conventional Weapons Which May Be Deemed to Be Excessively Injurious or to Have Indiscriminate Effects, White Paper No. 6, April 4, 2018.

[63] "Russia's Approaches to the Elaboration of a Working Definition," 2018.

[64] Bendett, 2018.

[65] Avramov interview, 2018.

[66] Igor Popov, "Военные Конфликты: Взгляд За Горизонт" ["Military Conflicts: A Look Beyond the Horizon"], Независимый Военный Обзор [*Independent Military Review*], n.d.

Threat Perception Could Drive Ethical Permissibility

Russia's wait-and-see approach to the ethics of AI is particularly unsettling when understood in the context of its current geopolitical worldview, which is characterized by an extreme wariness, approaching paranoia, about the threat of North Atlantic Treaty Organization (NATO) allies encroaching on its borders. In an analysis of Russia's military doctrine, Olga Oliker of the Center for Strategic and International Studies (CSIS) unpacked this idea further:

> Russia maintains, as it has in the past, that it will use military force only defensively, when other options have failed. But this is the doctrine of a state that sees a lot to defend against, even as its interests expand globally. While neither Moscow's overall goals nor the threats and dangers it faces have truly changed, it seems the Kremlin has grown more nervous that others are seeking ways to harm it, militarily and otherwise.[67]

In previous chapters we have alluded to a relationship between threat perception, risk tolerance, and ethical permissibility, and we shall show evidence of that relationship in U.S. public attitudes in Chapter 7. But put simply and directly here: as threat perception increases, so do risk tolerance and the ethical permissibility of using more autonomy. Consequently, wars perceived as defensive in nature—and thus highly threatening—may provide pretexts for a country to use AI without restraint.[68] It is quite possible that Russia would do just that. Worse still, if it becomes obvious that Russia cannot match the technology of the United States and China, Russia may compensate by using its own technology more ruthlessly.

Russia's Trust in Artificial Intelligence Could Be Eroded by Close Calls

Ethics aside, Russia's comfort with autonomy should not be overstated. The experts we interviewed expressed doubt that Russia would remove a human from the loop entirely, with one speculating that Russia does not trust its technology enough to do so.[69] Another interviewee made a similar remark, citing the 1983 Soviet nuclear false alarm as a historical precedent of retaining a human in the loop.

This incident occurred at a time of extremely heightened tensions in the Cold War. On September 26, 1983, the Soviet nuclear early warning system detected five Minuteman intercontinental ballistic missiles heading toward the Soviet Union from the United States. Soviet doctrine required the ranking officer at the early-warning command center to report the attack to his superiors, who would then issue a retaliatory nuclear strike. But Stanislav Petrov, the officer on duty, had reservations about the system's reliability, thinking it had been rushed into service. He found it odd that the system detected only five missiles; a nuclear first strike from the United

---

[67] Olga Oliker, "Russia's New Military Doctrine: Same as the Old Doctrine, Mostly," *Washington Post*, January 15, 2015.

[68] Avramov interview, 2018.

[69] One interviewee (name withheld upon request) mentioned that air defense could be an exception.

States was expected to be much larger. Petrov chose not to report the incident, deciding that the system had malfunctioned. This turned out to be the correct decision, and it likely prevented nuclear war and saved countless lives.[70]

Although the mistake was ultimately avoided, the Soviet Union got a taste of the potential pitfalls that can emerge from overtrusting machines. And of course, this incident occurred around the same time that the Politburo almost sparked nuclear war by obsessing over RYAN calculations, as discussed earlier in the chapter.

Russia may still want to make war a math problem, but it is possible that those close calls in the 1980s planted a seed of healthy skepticism in Russia's military culture about the dangers of automation bias.

Russia Prefers State Discretion to a Ban on Lethal Autonomous Weapon Systems

Although there are signs that Russia has reservations about autonomy, Moscow's statements expressing those concerns are often couched in vague terms that provide enough flexibility for Russia to change its tune at a later date. In particular, although the Russian delegation at the CCW says that maintaining a "due level of human control" over military weaponry is necessary, it resists efforts to develop criteria at the GGE to define what that means. Russia instead believes that the "forms and methods of such control should remain at the discretion of States." And in the same paper, Russia suggests that algorithms written by humans are sufficient as mechanisms of human control over systems with autonomous targeting and engagement capabilities.[71] This interpretation would permit the use of most, if not all, autonomous military systems in development or operation today.

The Russian delegation opposes a ban on LAWS, arguing that (1) LAWS is ill-defined and (2) the provisions of IHL are already a sufficient form of regulation.[72] Specifically, Russia has stressed the importance of Article 36 of Additional Protocol I from IHL, which calls for a legal review of new military weapon systems. Speaking for the delegation, Andrei Grebenschikov from the Russian Ministry of Foreign Affairs said that Russia "strictly adheres to these commitments," and he called on states that had not signed Additional Protocol I, such as the United States, to do so.[73] This view reiterates Russia's position that discretion should be left to the states, since there are no "mechanisms for international oversight or compliance with Article 36," nor are there "established international standards for undertaking weapon reviews."[74] Grebenschikov's comment was, however, slightly misdirected, since the United States already

---

[70] David Hoffman, "I Had a Funny Feeling in My Gut," *Washington Post*, February 10, 1999.

[71] "Russia's Approaches to the Elaboration of a Working Definition," 2018.

[72] "Russia's Approaches to the Elaboration of a Working Definition," 2018.

[73] Vadim Kozyulin, Tom Grant, Gilles Giac, Albert Efimov, Xinping Song Xian, and Mary Wareham, "Боевые Роботы: Угрозы Учтенные Или Непредвиденные?" ["Combat Robots: Recognized or Unexpected Threats?"], *Индекс Безопасности* [*Security Index*], No. 3–4, Vol. 22, p. 96.

[74] "Article 36 Reviews and Addressing Lethal Autonomous Weapons Systems," Article36, April 11–15, 2016, p. 2.

has a formal weapons review process that exists outside of the Article 36 construct and is far more transparent than Russia's.[75]

Conclusion

In the mid- to late 2000s, while the U.S. military started exploring how AI could alter warfare, Russia was still undergoing a number of military reforms and struggling to modernize old Soviet equipment. Only recently did it begin prioritizing the field of AI. Now Russia is dedicating resources toward developing this technology in a focused and unified effort, setting target goals, establishing interdepartmental standards, holding conferences, and building dedicated departments to foster innovation and integration of AI in the Russian military. There is a strong emphasis on robotics, but Russia is actively pursuing other areas such as defensive systems, decisionmaking and planning tools, EW, cyberwarfare, disinformation campaigns, and others.

Despite some successes, Russia's long-term prospects will be limited by structural, demographic, and cultural factors. It has a relatively small budget compared with its rivals, and it is experiencing an exodus of well-educated people from the country. Meanwhile, the population is both shrinking and aging. Traditionally, the military has put great trust in technology, but close calls with semiautonomous systems in the past could influence Russian leaders against fully unleashing autonomous systems today. Moreover, like China, Russia has a strong military tradition of centralized authority. It is difficult to imagine the Russian Army high command allowing subordinate units to use fully autonomous weapons, or even semiautonomous weapons, in scenarios in which mistakes that could result in escalation or other dangerous outcomes.

However, this would not stop Russia from using other forms of AI, particularly in the gray zone, where operations are smaller and there is less ethical, operational, and strategic risk. And as Vladimir Putin's political position weakens, he may become increasingly risk tolerant in these kinds of operations, as the recent poisonings of former Russian intelligence agents in the United Kingdom seem to suggest.[76] More seriously, while Russian military leaders might be reluctant to authorize unrestricted autonomy in conventional warfare, if a conflict were to escalate to a point that Moscow felt an existential threat on its own soil or in Russia's near abroad, concerns about autonomy would likely vanish. Russian leaders would do whatever they believe is necessary and justify their actions on defensive grounds. In a similar manner, if the United States or China

---

[75] Brian Rappert, Richard Moyes, Anna Crowe, and Thomas Nash, "The Roles of Civil Society in the Development of Standards Around New Weapons and Other Technologies of Warfare," *International Review of the Red Cross*, Vol. 94, No. 886, Summer 2012, p. 781.

[76] This presumes that those acts were directed, or at least authorized, by the Kremlin. They might have been actions directed by officials at lower levels in the Russian intelligence bureaucracy.

achieved a decisive superiority in AI, Russia might be inclined to use its own AI-equipped technology more aggressively to compensate.

Managing strategic concerns is Russia's main priority. Moscow may feel that ethics is a luxury it cannot afford.

7. Assessment of U.S. Public Attitudes Regarding Military Artificial Intelligence

A variety of stakeholders and advocates have expressed concern about the ethical implications of military AI. But it is not clear how broadly these apprehensions resonate outside of the most vocal groups. In order to understand the U.S. general public's views of military AI, we developed and executed a survey concerning near-term AI systems in specific military applications.[1] This chapter discusses the importance of assessing the public's attitudes, describes prior public surveys regarding military AI, and considers the results of the survey we administered.

Importance of Public Perception

Views regarding the ethical permissibility of military action and the role of technology are not just the domain of academic ethicists or government lawyers. Members of the general public also have perspectives about the ethical principles that apply in war, and they might have views about what types of military technologies should be developed and when and how leaders should employ them.

Public perception of these issues is important in part because it might legitimize leaders' decisions to use certain weapons, or it might urge caution. The growth of the internet and the use of smartphone cameras, social media, and other open-source data streams have rapidly increased public access to information about military operations. This ever-growing access allows citizens to increasingly scrutinize operations conducted on their behalf. On the one hand, the public might express opposition to specific military actions and pressure the government to change course. On the other hand, the public might recognize the necessity or righteousness of the actions and rally behind the decision, thereby providing public legitimation and endorsement. In this way, the public serves as a source of ethical reflection and moral integrity that can translate into political action and national-level decisionmaking. Especially in democracies, greater understanding of the public's view can help inform leaders and enable them to plan operations that are consistent with accepted values and norms. Not only do leaders need to know the public's ethical thresholds for military technology, but they also need to understand what factors influence those views.

Another consideration underscores the importance of public assessment of military conduct. This one is based on the Martens Clause, an important passage in LOAC. First articulated in the preamble in the 1899 Hague Convention, this clause later appeared in the Geneva Conventions

---

[1] We conducted this survey using Amazon's MTurk, a crowd-sourcing platform. For more on the methods we employed and their potential drawbacks, see Appendix B.

Additional Protocol 1 and elsewhere. It articulates a general principle intended to minimize harms in armed conflicts.[2] The Martens Clause states:

> In cases not covered by this Protocol or by other international agreements, civilians and combatants remain under the protection and authority of the principles of international law derived from established custom, from the principles of humanity and from *the dictates of public conscience*.[3]

The Martens Clause specifies a principle that, in cases not codified in LOAC, belligerents are legally mandated to conduct hostilities in accordance with the dictates of "public conscience." Although there are varied legal interpretations of the clause and open questions about the concept of public conscience, the Martens Clause is widely interpreted to stress that the public's views of military conduct should provide some guidance on the legal deliberations of governments. As explained by the International Committee for the Red Cross, the Martens Clause "is a safety net for humanity."[4] Thus, public perception is important not only from a standpoint of political legitimacy but also from the standpoint of international law.

Advocacy groups have sought to galvanize public attention against the development of autonomous weapons and to support an international treaty that would restrict their use. Some critics have argued that the Martens Clause prohibits the use of autonomous weapons because public sentiment is opposed to them. However, only limited data available regarding U.S. public opinion are available, and thus in order to assess the claim of these critics, it is important to better understand the actual views of the public.

Prior Surveys

We have identified two prior surveys relevant to assessing U.S. public opinion regarding autonomous weapons. In 2013, Charli Carpenter conducted a two-question survey of 1,000 random American respondents aged 18 or older via YouGov America.[5] The first question asked respondents, "How do you feel about the trend toward using completely autonomous [robotic weapons/lethal robots] in war?"[6] The survey found that 55 percent of respondents either somewhat opposed or strongly opposed them. Interestingly, among those respondents who were currently in the military, an even higher percentage (73 percent) opposed them. The second

---

[2] Ticehurst, 1997.

[3] International Committee of the Red Cross, *Protocols Additional to the Geneva Convention of 12 August 1949*, Art. 1, Para. 2 (emphasis added).

[4] International Committee of the Red Cross, "Ethics and Autonomous Weapon Systems: An Ethical Basis for Human Control?" April 2018.

[5] Charli Carpenter, "How Do Americans Feel About Fully Autonomous Weapons?" Duck of Minerva, June 19, 2013.

[6] The wording of the question alternated between "robotic weapons" and "lethal robots" to determine whether those word choices affected responses.

question asked respondents the extent to which they support or oppose a global treaty requiring human involvement in all decisions to take human life. In response, 53 percent indicated they supported a global treaty. Again, the number was even higher (60 percent) among those in active military service.

A second survey by Michael Horowitz in 2016 consisted of four questions administered to 1,043 random American respondents using Amazon's Mechanical Turk (MTurk) platform. The survey sought to explore the factors that influence respondents' attitudes about autonomous weapon systems. The survey found that "fear of other countries or non-state actors developing these weapons makes the public significantly more supportive of developing them, as does a perception that they are necessary to protect U.S. troops from attack."[7] In this way, respondents' views about the ethical permissibility of autonomous weapon systems was contextual, based on these factors.

These surveys provide important information for understanding the public's views, but both involved generic questions about autonomous weapons and did not inquire about specific military applications of AI. In order to fill this gap, we designed and executed a more expansive survey to get a more detailed understanding of the public's ethical assessment of a range of military AI applications.

Results of the Survey of U.S. Public Attitudes

This section reports the results of our survey on U.S. public attitudes regarding the use of military AI using Amazon's MTurk crowdsourcing platform.[8] As we shall discuss, the survey responses suggest that Americans are concerned about the risks and ethical implications of using this technology and certain applications of it in particular. However, the results also indicate that Americans appreciate the potential benefits that these capabilities might provide, and a significant majority of citizens supports continued investments in them. Public support for the employment of military AI systems depends on context: Defensive systems are more acceptable than offensive systems; using machines to attack machines is more acceptable than using them to attack people; and keeping a human in the loop in the use of lethal force is important. Finally, the respondents were surprisingly supportive of using AI to improve ISR processing and decisionmaking, but sharply rejected using it to falsely attack the reputations of foreign leaders.[9]

---

[7] Michael C. Horowitz, "Public Opinion and the Politics of the Killer Robots Debate," *Research and Politics*, January–March 2016, p. 2. The term "more supportive" in the quotation refers to respondents' increased support for developing autonomous weapons when they fear other countries are developing them or when they perceive they are necessary to protect U.S. troops from attack relative to their support without this fear or perception.

[8] See Appendix B for details related to how we used the MTurk system.

[9] This chapter provides an interpretation of the survey results using select excerpts from survey responses. Readers are encouraged to review the full set of survey responses in Appendix B. There, we also explain the methods employed in developing and administering the survey, provide the demographics of respondents, and do a series of statistical analyses of the results.

Concerns About the Employment of Military Artificial Intelligence

The survey responses suggest that the U.S. public has significant misgivings about military applications of AI. These apprehensions revolve around impressions that autonomous weapons lack accountability, violate human dignity, and inappropriately remove human emotion from war. Respondents also believed that the development of these weapons would make war more likely. As a result, a majority felt that the United States should work with other countries to ban them.

Accountability, Human Dignity, and Human Emotions Matter

As Figure 7.1 suggests, survey respondents expressed concern about the accountability of autonomous weapons. Approximately 57 percent of respondents believe that autonomous weapons are ethically prohibited because they cannot be held accountable for wrongful actions.

Figure 7.1. Public Concerns About the Lack of Accountability of Autonomous Weapons

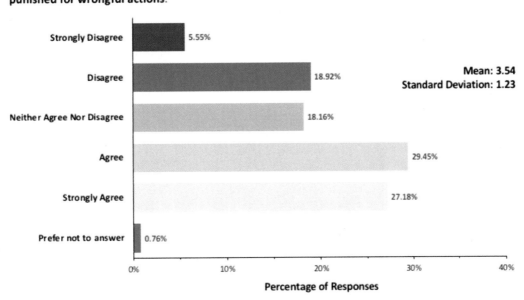

As discussed in Chapter 2, some experts have expressed the opinion that one of the benefits of autonomous weapons is that it will remove human emotion from war, helping commanders make decisions more objectively and with a clearer conscience. However, as Figure 7.3 indicates, 53 percent of the respondents to our survey felt that human emotion should not be removed from war. Only 29 percent of them felt that doing so would be beneficial.

Similarly, Figure 7.2 indicates that a significant percentage of the public feels that people being killed by autonomous weapons would violate the dignity of human life. Although the percentage is less than a majority, the 46 percent of people expressing this view is significantly greater than the percentage of those who disagree with it. The 24 percent who neither agreed nor disagreed suggests there is a degree of ambivalence about this issue.

Figure 7.2. Public Sentiment Regarding Autonomous Weapons and Human Dignity

Autonomous weapons are ethically prohibited because they violate the dignity of human life.

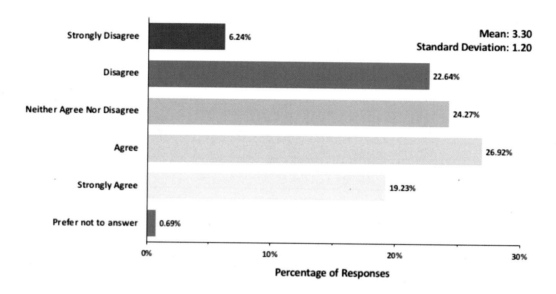

Figure 7.3. Public Sentiment on Human Emotion and War

Removing human emotions from decisions in war is beneficial.

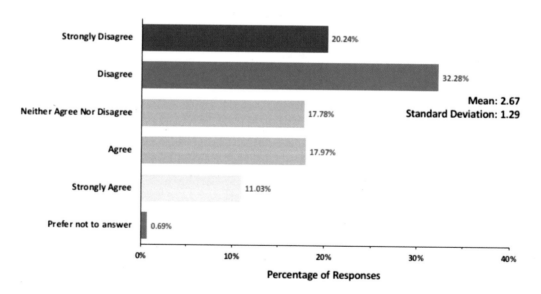

Risks of War and the Desire to Ban Autonomous Weapons

Another concern that many survey responses reflected was the potential effect that the development of autonomous weapon might have on the occurrence of war. As Figure 7.4 indicates, over 53 percent of survey respondents felt that developing autonomous weapons will make wars more likely.

Figure 7.4. Public Opinions About Autonomous Weapons and the Likelihood of War

The development of autonomous weapons will make the occurrence of wars more likely.

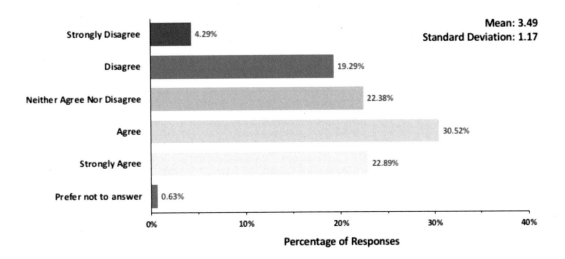

Perhaps owing to this concern and those above, a majority of survey respondents felt that the United States should work with other countries to ban autonomous weapons. As Figure 7.5 indicates, 53 percent of respondents agreed with this position. Only 22 percent of respondents disagreed.

Figure 7.5. Public Opinion on the Need for an International Ban on Autonomous Weapons

The United States should work with other countries to ban autonomous weapons.

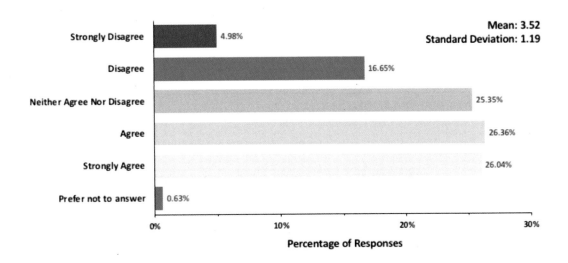

These opinions were held more strongly among Democrats and women than Republicans and men. In keeping with Carpenter's survey findings, respondents with military experience felt more strongly than those without military experience that autonomous weapons should be banned.[10]

Public Recognition of the Benefits of Military Artificial Intelligence

Despite the concerns discussed above, there were clear indications in the survey responses that the U.S. public recognizes the potential benefits of military AI. For instance, as Figure 7.6 illustrates, a large majority (63 percent) of respondents believe that autonomous weapons will be more accurate and precise than human operators.

Figure 7.6. Public Belief That Autonomous Weapons Will Be More Precise Than Humans

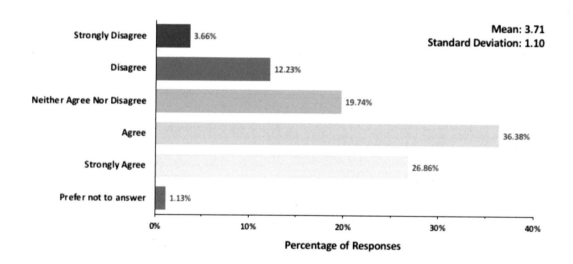

Autonomous weapons will be more accurate and precise than humans.

And as we shall discuss below, the respondents were surprisingly supportive of military applications of AI that did not involve autonomous weapons and even those with a degree of autonomy, but with a human operator in the loop.

Perhaps owing to these considerations, a significant majority of respondents believe the United States should continue to invest in AI for military use. Figure 7.7 shows that 63 percent agree with such investments, whereas only about 16 percent oppose them.

Interestingly, in this case, respondents with military experience agreed more strongly than others that autonomous weapons will be more accurate and precise than human operators. They also supported continued investment in AI for military use more strongly. And surprisingly,

---

[10] See Appendix B for a more detailed demographic analysis.

Figure 7.7. Public Support for Continued U.S. Investment in Military Artificial Intelligence

It is ethically permissible for the U.S. military to continue to invest in artificial intelligence technology for military use.

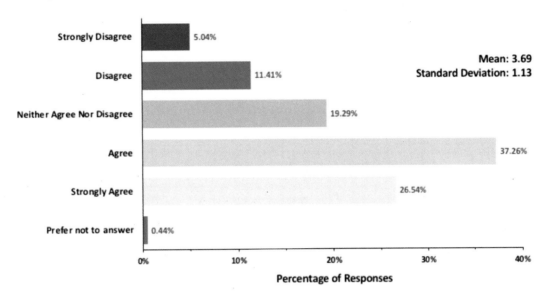

almost half (49 percent) of respondents who agreed with the statement "War is always wrong" supported continued investment in these technologies.

Public Acceptance of Military Artificial Intelligence Depends on Context and Degree of Human Control

As reported above, survey responses expressed ethical concern and anxiety about the risks of autonomous weapons, yet acknowledged their potential benefits and supported continued investment in military AI. This may seem contradictory, but answers to other survey questions might help explain these results. In short, the survey suggests that public acceptance of military applications of AI will depend on the context in which it is employed and whether human operators have the appropriate degree of control over these systems in each particular situation.

Autonomous Weapons in Offensive Operations

An example of the contextual nature of survey responses can be seen in respondents' discomfort with the use of autonomous weapons in offensive operations. For instance, as Figure 7.8 illustrates, respondents were strongly against using missiles that search for and destroy enemy targets without human authorization. Almost 70 percent of respondents objected to this, and only about 17 percent of them agreed that it was acceptable.

Figure 7.8. Autonomous Missiles in Offensive Operations Without Human Authorization

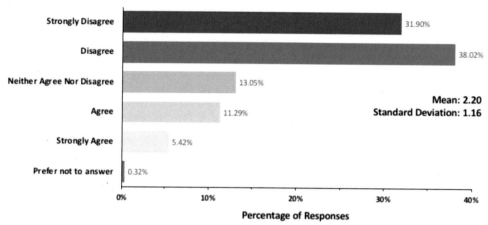

It is ethically permissible for the U.S. military to use missiles that autonomously search for and destroy enemy targets in warzones without human authorization.

Objections were even stronger when the statement specified that targets would be in close proximity to civilians. Then 72 percent objected, with support for it dropping to 15 percent.

However, if the autonomous missiles were not permitted to strike targets without human authorization, the responses were very different. Figure 7.9 shows that a large majority of respondents felt that autonomous missiles should be allowed to search for and destroy enemy targets only if they have human authorization. In fact, the percentages are essentially reversed from the previous chart.

Figure 7.9. Autonomous Missiles in Offensive Operations With Human Authorization

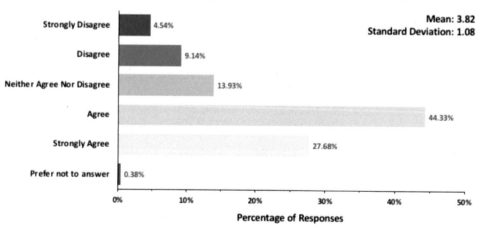

It is ethically permissible for the U.S. military to use missiles that autonomously search for and destroy enemy targets in warzones only if the missiles have human authorization.

So this suggests that an important factor in the public's acceptance of using autonomous weapons in offensive operations is whether human operators have control over them when they strike. As we shall see, this factor, while still important, becomes less so when conducting defensive operations or when the enemy is winning a battle.

Autonomous Weapons in Defensive Operations

Survey respondents were much more accepting of autonomous weapons when U.S. forces were being attacked or when the United States was losing a battle. For instance, Figure 7.10 shows that they strongly agreed with letting a swarm of autonomous drones attack enemy drones if they were attacking U.S. troops.

Figure 7.10. Defensive Use of Autonomous Drones Against Enemy Drones

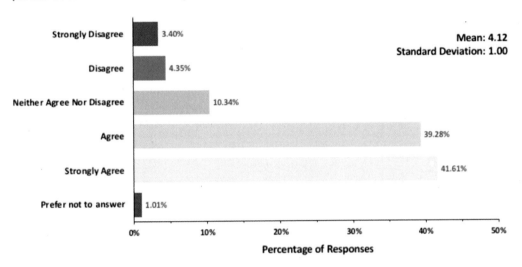

It is ethically permissible for the U.S. military to use a swarm of armed autonomous drones to protect US soldiers from an enemy autonomous drone swarm that is attacking.

And as Figure 7.11 shows, this support was almost as strong for letting a U.S. swarm of autonomous drones preemptively attack a swarm of enemy drones.

Figure 7.11. Preemptive Use of Autonomous Drones Against Enemy Drones

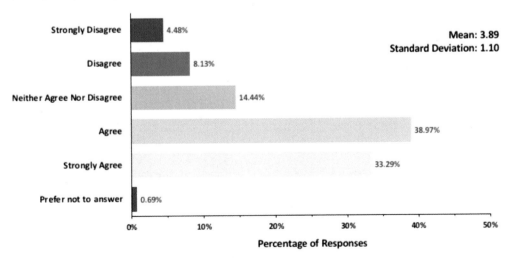

It is ethically permissible for the U.S. military to use a swarm of armed autonomous drones to preemptively destroy an enemy autonomous drone swarm.

Of course, the operative variable here might be that the drones are not taking human life in these scenarios; they are just attacking enemy machines.

Indeed, when we asked respondents if they agreed with using drone swarms against enemy combatants, the results were somewhat different, as Figure 7.12 illustrates.

Figure 7.12. Use of Autonomous Drones to Attack Enemy Combatants

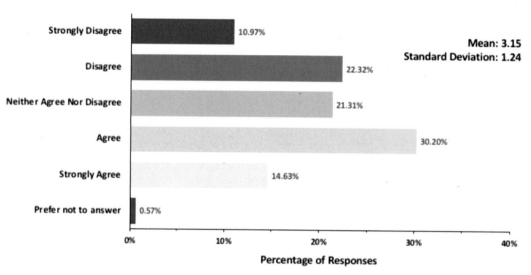

As the figure indicates, less than a majority of survey respondents agree with using autonomous drones to attack enemy combatants, although, at about 45 percent, support for this action was still greater than the 33 percent who opposed it. All of this suggests that the U.S. public is more supportive of using autonomous weapons in defensive roles than in offensive roles, and human control over their employment is important in all cases in which those weapons might take human life.

Tendencies Toward Escalation

Despite these constraints, certain survey responses indicate that they might be relaxed in certain situations, such as if the United States were losing a battle. For instance, as Figure 7.13 indicates, respondents were more willing to accept the use of autonomous weapons in efforts to stave off defeat, even if the enemy is not using those capabilities.

Figure 7.13. Use of Autonomous Weapons to Avoid Defeat When the Enemy Is Not Using Autonomous Weapons

It is ethically permissible for the U.S. military to use autonomous weapons when the enemy is winning a battle without autonomous weapons.

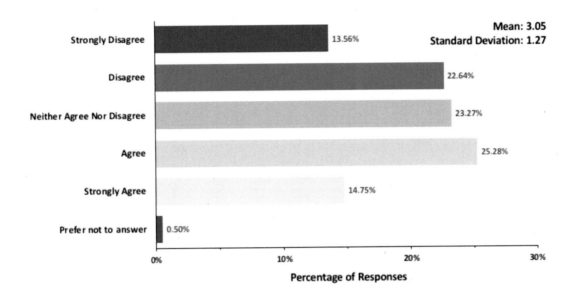

In this situation, 40 percent of respondents agree with using autonomous weapons and 36 percent disagreed, with 23 percent undecided (those who neither agreed nor disagreed). While 40 percent in support is less than a majority, it is clearly stronger than when respondents considered using autonomous weapons against enemy targets without human authorization, in which case 70 percent of respondents objected. The ambivalence reflected in the response to this survey question suggests respondents may be torn between their ethical concerns about autonomous weapons and their willingness to let U.S. forces do what is needed to avoid defeat. Forty percent of them chose the latter, reflecting a natural human tendency to escalate to avoid losing a fight.

Figure 7.14, showing the acceptance of using autonomous weapons when an enemy is using autonomous weapons, introduces another escalation dynamic.

Figure 7.14. Use of Autonomous Weapons to Avoid Defeat When the Enemy Is Using Autonomous Weapons

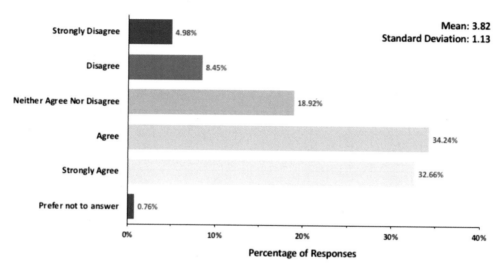

It is ethically permissible for the U.S. military to use autonomous weapons when the enemy is winning a battle with autonomous weapons.

Mean: 3.82
Standard Deviation: 1.13

Strongly Disagree — 4.98%
Disagree — 8.45%
Neither Agree Nor Disagree — 18.92%
Agree — 34.24%
Strongly Agree — 32.66%
Prefer not to answer — 0.76%

Percentage of Responses

In this situation, a very strong majority of respondents supported the use of autonomous weapons: 67 percent agreed and only 13 percent disagreed. This result illustrates a fundamental dynamic of escalation: When one side crosses a critical escalation threshold—in this case, the use of autonomous weapons—the other side feels more justified in doing so as well. This suggests that, given the typical ease of switching weapons from semiautonomous to fully autonomous modes, there would be significant risks of escalation should we find ourselves at war and both sides are armed with autonomous weapons.

Public Acceptance of Autonomy at Rest

In Chapter 2, we talked about the differences between autonomy-in-motion, machines taking physical action, and autonomy at rest, software processing that supports human decisionmaking but does not directly act in the physical world. There, we cautioned that although autonomy at rest might be less risky than autonomy in motion, serious risks are still present, because the application could advise decisionmakers to take lethal action or do other dangerous things without operators or leaders being able to evaluate the quality of those recommendations in a timely manner. It seems that respondents to our survey either do not appreciate these risks, or do not feel they are severe enough to object to certain applications of military AI. They found some applications of autonomy at rest very acceptable. Yet, they found others highly objectionable on ethical grounds.

Highly Accepted Forms of Autonomy at Rest

Figure 7.15 illustrates the survey respondents' high acceptance of a military application of autonomy at rest.

112

Figure 7.15. Use of Military Artificial Intelligence to Identify Enemy Targets

It is ethically permissible for the U.S. military to use a computer program to analyze data to identify the location of enemy targets.

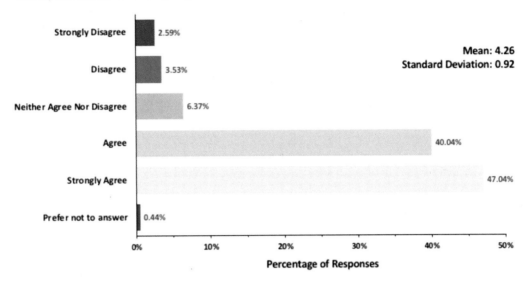

As the figure indicates, an overwhelming majority of respondents (87 percent) agreed with using a computer program to analyze data to identify the location of enemy targets. Only 6 percent of respondents had problems with this. This may reflect a degree of "automation bias," a belief that AI systems are more reliable and trustworthy than humans.

Similarly, as Figure 7.16 indicates, our survey respondents were also strongly supportive of allowing computer programs to advise commanders on how to attack enemy targets.

Figure 7.16. Use of Military Artificial Intelligence to Advise Commanders on How to Attack Enemy Targets

It is ethically permissible for the U.S. military to use a computer program to make recommendations to a military commander on how to attack targets.

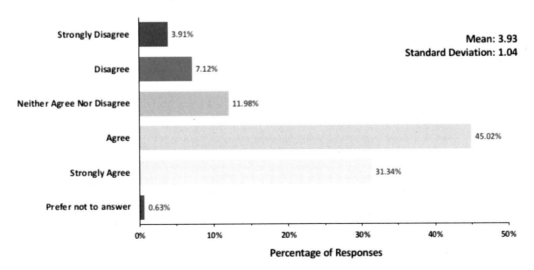

With 58 percent of respondents agreeing with this application of military AI, versus 11 percent disagreeing with it, this form of autonomy at rest appears to have the faith and confidence of the U.S. public.

In fact, survey support for these forms of military AI remained strong even for applications that might venture into controversial areas, such as using facial recognition or other forms of biometric analysis to identify enemy combatants. Figure 7.17 illustrates this point.

Figure 7.17. Use of Biometric Analysis at Military Checkpoints to Identify Enemy Combatants

It is ethically permissible for the U.S. military to use a robot with facial recognition or other biometric analysis at a military checkpoint to identify and report enemy combatants.

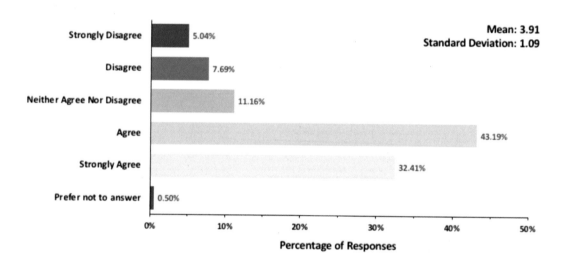

With over 65 percent of respondents supporting this application and less than 13 percent opposing it, it would appear that the U.S. public would be comfortable with using military AI in this fashion.

And surprisingly, the respondents were even accepting of using a similar application that crosses the line into autonomy in motion—in this case, allowing robots to identify and subdue enemy combatants—as Figure 7.18 reveals.

As the figure illustrates, 62 percent of respondents still agreed with using military AI in this manner, while less than 20 percent of them disagreed with it. One would wonder what percentage of them would have agreed had we worded the statement differently, stipulating that the robot would use lethal force, if necessary, to subdue the suspect. Presumably, considering responses to previous statements, there would be less support for autonomous action, given that human life would be at stake. Nevertheless, responses to this series of survey statements suggest that the U.S. public is strongly supportive of most military applications of AI focused exclusively, or even primarily, on autonomy at rest.

114

Figure 7.18. Use of Biometric Analysis and Robotics to Identify and Subdue Enemy Combatants

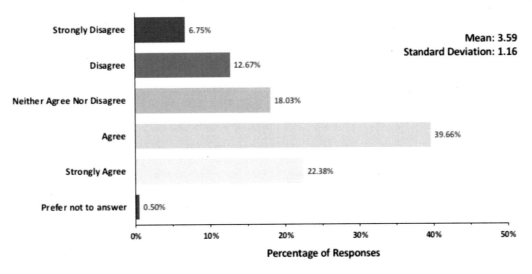

It is ethically permissible for the U.S. military to use a robot with facial recognition or other biometric analysis at a US military checkpoint to identify and subdue enemy combatants.

Strongly Disagree — 6.75%
Disagree — 12.67%
Neither Agree Nor Disagree — 18.03%
Agree — 39.66%
Strongly Agree — 22.38%
Prefer not to answer — 0.50%

Mean: 3.59
Standard Deviation: 1.16

Percentage of Responses

So what applications of autonomy at rest did respondents object to?

Less Accepted Applications of Autonomy at Rest

Although responses to our survey showed very strong support for some military applications of autonomy at rest, they were less supportive of others, and some they strongly opposed. For instance, Figure 7.19 indicates that many respondents were not comfortable with exploiting a vulnerability in commercial software for military advantage, versus reporting the problem to the company.

Figure 7.19. Concerns About Exploiting Vulnerabilities in Commercial Software

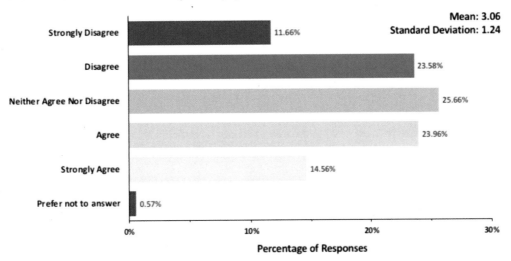

It is ethically permissible for the U.S. military to exploit new vulnerabilities in commercially available software to attack enemy military systems rather than notify the company of the bug.

Strongly Disagree — 11.66%
Disagree — 23.58%
Neither Agree Nor Disagree — 25.66%
Agree — 23.96%
Strongly Agree — 14.56%
Prefer not to answer — 0.57%

Mean: 3.06
Standard Deviation: 1.24

Percentage of Responses

115

Although the 39 percent of respondents who agreed with doing this edged out the 35 percent of them who disagreed, this split, along with the large group of undecided respondents (the 26 percent who neither agreed nor disagreed) suggests there is significant ambivalence and ethical discomfort about this kind of activity.

In contrast, respondents overwhelmingly rejected using AI to falsely attack a foreign leader's reputation, as Figure 7.20 indicates.

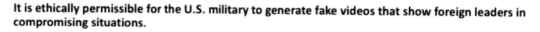

Figure 7.20. Use of Artificial Intelligence to Generate Fake Videos to Show Foreign Leaders in Compromising Situations

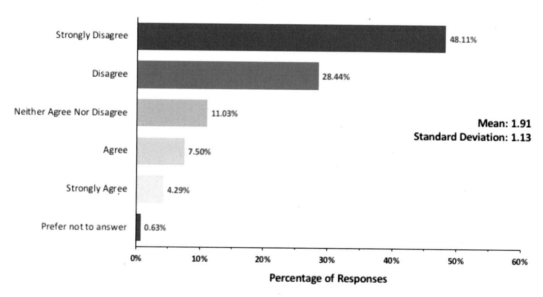

It is ethically permissible for the U.S. military to generate fake videos that show foreign leaders in compromising situations.

Over 76 percent of respondents objected to this kind of operation, with nearly half of all respondents strongly disagreeing with it. In fact, this action was the one respondents found the most ethically objectionable in the entire survey, even more so than using autonomous missiles to attack targets in close proximity to civilians, which received 72 percent disagreement.[11]

Conclusion

The results of this survey not only corroborate the findings of the Carpenter and Horowitz surveys but also provide a much richer understanding of public attitudes regarding military AI. Our findings suggest that the U.S. public has serious reservations about certain military applications of AI, but their objections are stronger in certain situations than in others. Large

---

[11] Both of these questions had a 2-percent margin of error at the 95 percent confidence level.

majorities of survey respondents objected to using autonomous weapons to take human life without a human operator in control. But these concerns were strongest when the weapons were used in offensive operations, more relaxed when they were used defensively, and nearly disappeared when they were used only to destroy other autonomous weapons. Most significantly, despite the respondents' apparent abhorrence to the use of autonomous weapons to kill people, they were more willing to do so if U.S. forces were losing a battle, especially if the enemy was using autonomous weapons. Assuming this aspect of human nature is operative in all countries, this suggests that wars will likely be prone to escalation if both sides are armed with autonomous weapons.

Interestingly, the results of this survey indicate that the U.S. public has few qualms at this point about using military applications of autonomy at rest in functions such as ISR processing or decision support. This would suggest that few U.S. citizens have anxieties about Air Force efforts to harness AI for enhanced ISR processing, such as in Project Maven. However, respondents were strongly averse to using AI to falsely attack the reputations of enemy leaders.

These observations inform our findings and recommendations, which we address in the next chapter.

8. Findings and Recommendations

According to Carl von Clausewitz, the "nature" of war—*"an act of force to compel the enemy to do our will"*—will always remain the same.[1] However, the "character" of war, the way war is conducted in a specific time and place, will be shaped by contextual factors of the historical moment.[2] Just as technologies, such as gunpowder and aircraft, have fundamentally altered the character of war in the past, AI has the potential to change how wars are fought in the future. That future may be closer than many people realize.

AI is not just a single technology. It is a class of technologies that can be integrated into a range of applications. These applications can, among other things, increase the speed of decisions and enable new forms of military analysis and combat power. AI potentially portends a dramatic evolution, perhaps even a transformation, in the character of war.

However, the application of AI in war raises new and complex ethical questions regarding its role vis-à-vis the role of human warfighters. Such questions include whether AI systems can comply with humanitarian principles, whether they will be sufficiently reliable and predictable, and what effects they will have on escalation and stability.

This report has reviewed military AI developments in the United States, China, and Russia and examined the ethical questions surrounding the potential employment of these technologies. This concluding chapter summarizes our key findings and identifies five specific recommendations for the Air Force to manage the ethical implications of AI in war.

Findings

There Is No Consensus on an Artificial Intelligence Development Timeline, but Experts Agree That There Will Likely Be a Steady Increase in the Integration of Artificial Intelligence in Military Systems

There is broad acknowledgment that AI technologies have developed rapidly and are being integrated into an increasing number of applications; however, there is no consensus among experts about the timeline for further AI development. Some experts focus their attention on near-term applications, those expected to be operational in the next few years, while others analyze implications of "superintelligent" systems that might be decades away, or never

---

[1] Carl von Clausewitz, *On War*, ed. and trans. Michael Howard and Peter Paret, Princeton, N.J.: Princeton University Press, 1976, p. 75 (emphasis in original).

[2] See Christopher Mewett, "Understanding War's Enduring Nature Alongside Its Changing Character," *War on the Rocks*, January 21, 2014.

reached.[3] Even considering the near- and midterm military applications of AI, experts do not agree on which ones will progress most rapidly. However, they all recognize that the various forms of AI represent important technological developments with serious ramifications for a wide range of warfighting applications.

Different development timelines of military AI will have different implications for ethical issues in war. If AI develops at a rapid pace with major advances in object recognition, decision support, cybersecurity, and other key warfighting applications, then military establishments around the world will hasten to integrate these technologies widely. This could foretell a rapid evolution in the ways wars are fought, perhaps diminishing or altering the role of human agency in war. In this scenario, autonomous systems will play an outsized role in warfare that will increase the speed of action potentially beyond human ability to direct or constrain them. If this rapid development occurs, the ethical, operational, and strategic risks that this report identifies will be of utmost concern, and stakeholders will need to move quickly to mitigate the most extreme risks.

On an alternative trajectory, the field may encounter technological roadblocks that could greatly diminish its current level of enthusiasm and funding. As has happened in the past, progress could soon plateau and be followed by a long period of stagnation, or an "AI winter." In this scenario, ethical considerations surrounding war will remain much the same as today. While those ethical questions are by no means trivial, they will not be further complicated by machines that operate autonomously, since humans will still remain in control of warfighting functions.

That said, the most likely scenario is one in which the development of military AI will continue along a steady trajectory, with advances in a range of applications that military establishments will seek to exploit. In this path, AI will present new ethical questions in war but at a pace at which deliberate attention can potentially mitigate the most extreme risks. This will allow time for cautious analysis to determine the best ways to ensure a meaningful human role in control over military AI systems.

Military Artificial Intelligence Offers a Wide Range of Potential Benefits, and the United States Faces Significant International Competition in This Field

Chapter 2 provided an overview of the potential military benefits of AI, including faster and better decisionmaking, improved ISR and precision targeting, mitigation of manpower issues, and improvements in cyber defense. Hoping to harness these benefits to regain military advantages eroded by the post–Cold War proliferation of advanced military capabilities, DoD is investing heavily in these technologies. However, private companies are leading the research on the most innovative applications, and government institutions will need to leverage these technology firms for their own acquisition and development. Moreover, the United States is not the only nation

---

[3] Nick Bostrom, *Superintelligence: Paths, Dangers, Strategies,* New York: Oxford University Press, 2014.

whose military establishment is seeking to benefit from AI. Unlike in some past technological developments, the United States will not have a monopoly, or even a first-mover advantage in this competition.

China is aggressively pursuing militarized AI technologies, and its developers may have certain advantages over their U.S. counterparts. Potential Chinese advantages include a top-down strategic approach that emphasizes civil-military fusion, access to vast amounts of data for algorithm training, and a potential willingness to press into risky and arguably unethical technology applications. However, China also faces limitations in its AI advancement, including a dearth of AI experts and technicians.

Russia is also seeking to rapidly develop military AI. Political and structural factors in Russia both help and hurt its ability to integrate AI into military applications. Although Russia's private sector is less diversified and developed than that of the United States, like China's, it is potentially more responsive to military requirements. Suffering from Western sanctions and hampered by an inflexible, centrally controlled defense-industrial base, Russia is in a position of relative economic weakness, but this position encourages it to develop AI applications aggressively and employ them more recklessly. Indeed, Russia is already using AI technologies in support of its military efforts. They provide important capabilities in its hybrid, gray-zone, and information warfare campaigns.

The United States also faces significant risks that military AI could proliferate to other state and nonstate actors. Some tools, such as offensive cyber capabilities, feature relatively low development costs and easy reuse once they are loose in the environment. Actors such as North Korea and criminal groups have already been able to harness some of these capabilities for their malicious purposes. As the cost of AI capabilities decrease, applications could proliferate to a broad range of actors with potentially lethal consequences.

The Development of Military Artificial Intelligence Presents a Range of Risks That Need to Be Addressed

The U.S. public, key researchers and technologists, and elements of the U.S. private sector have all raised concerns about the risks associated with military AI. We have surveyed these risks and sorted them into three categories: ethical, operational, and strategic. Each of these risk categories presents challenges that need to be addressed, not just by the United States but by all actors involved in the development and employment of military AI.

Ethical risks are particularly important from a humanitarian standpoint. States are obligated to abide by the provisions of IHL that seek to protect innocent civilians from the violence and abuses of war. Autonomous weapons raise fundamental questions about the protection of human dignity and whom to hold responsible for harmful action. Such ethical questions are important not only from a humanitarian standpoint; governments also need to address them in terms of how their military forces will apply AI in order to earn the trust of their public- and private-sector stakeholders.

Operational risks, such as those related to the reliability, fragility, and security of AI systems, also need to be taken seriously. These operational risks raise fundamental questions about whether military AI systems will function according to the intent of military commanders and operators. Addressing these risks through testing and evaluation and by building in safety procedures and appropriate rules of engagement is crucial for U.S. military forces and those of other countries.

Strategic risks, including the risks that AI will increase the likelihood of war, escalate ongoing conflicts, and proliferate to malicious actors, are also important to the international community. These risks need to be better understood to ensure that AI systems intended to promote national security will not decrease it instead, or even undermine global stability.

China and Russia are not immune to these risks, although they may be less sensitive to some than others. For instance, China has proposed a ban on LAWS, which Russia and the United States do not support—but Beijing's proposed ban defines LAWS so narrowly that it probably would not constrain China's development or use of these weapons even if the international community accepted it. This leads us to question whether Beijing's professed concerns about human dignity and moral responsibility are genuine. And China and Russia are clearly less sensitive to some other ethical concerns, such as their citizens' rights to privacy.

On the other hand, there is reason to believe Beijing and Moscow genuinely do care about the operational and strategic risks entailed in military AI. No military or political leader wants lethal weapons that are unreliable, can be hacked, or might exhibit unpredictable emergent behaviors. Nor do any national leaders want their military commanders to be advised by decision support systems that might recommend actions that are insensitive to escalation thresholds and thereby risk stability in a crisis or escalation in war. In fact, these concerns might be even greater in China and Russia than in some other countries, given their political and strategic cultures, which emphasize centralized control.

International Competition Could Encourage Countries to Rush the Development of Military Artificial Intelligence Without Sufficient Attention to Safety, Reliability, and Humanitarian Consequences

Competition between states creates incentives for them to rapidly develop and integrate AI technology into military applications. However, there is a risk that rapid development will come at the cost of safety, reliability, or compliance with humanitarian principles. Although some states have sought to develop effective legal reviews, testing and evaluation regimes, and other safeguards for military AI, many states do not have such restrictions or have not publicly explained how they will ensure that risks are mitigated.

Since DoD's 2012 release of Directive 3000.09, the United States has had the most advanced policy on the regulation of autonomous weapons that is publicly available. The directive articulates a standard of "appropriate human judgment" in development and use of autonomous weapon systems, and requires training and other policy guidelines to ensure that autonomous weapons can be used reliably and safely. Other countries, however, have not been as transparent about military AI policy. Russia and China have not publicly articulated restraints; nor have they explained how their legal review process ensures compliance with LOAC. It is also not clear that the United States is aligned with NATO members or other allies regarding policies that apply to the development and use of autonomous weapon systems.

International competition in the development of military AI could escalate into a full-blown arms race. The lack of international consensus on norms of responsible development and use creates risks that states will have an incentive to rapidly acquire and integrate military AI without putting appropriate policies in place. Such an environment could generate ever-increasing pressure to quickly identify and develop new military AI applications without sufficient precaution to ensure they are safe and reliable. This situation could result in a "race to the bottom," ultimately threatening the ability of humans to exercise agency over military AI systems. Such an outcome would have serious ramifications for the entire international community.

The U.S. Public Appears to Support the Department of Defense's Continued Investment in Military Artificial Intelligence, but the Public's Acceptance of Risk Varies by Context

AI is already integrated into countless everyday civilian applications, such as online search functions, recommendation algorithms, navigation systems, and so forth; thus, the public is increasingly comfortable with AI. The results of our survey suggest that the public also supports DoD's investment in military AI applications, and citizens agree with DoD's contention that autonomous weapon systems might be more accurate and precise than humans.

However, the results of our survey also indicate that the public is concerned about the ethical risks that military AI poses for accountability and human dignity. Similarly, the public appears to hold strong convictions about the importance of human involvement in the use of autonomous weapons and that an operator should be required to authorize attacks that take human life. The survey indicates that the public has different degrees of support and concern for different military AI applications. For instance, the survey found that public was less concerned about decision support systems (such as Project Maven) than other uses of military AI applications such as generating fake videos.

Public support for military AI depends in part on contextual factors such as whether the adversary is using autonomous weapons or whether the system is necessary for self-defense. Thus, ethical risks from the public's perspective are not always bright lines but vary according to the threat landscape and other considerations.

Despite Ongoing UN Discussions, an International Ban on or Other Regulation of Artificial Intelligence in Military Applications Is Not Likely in the Near Term

The UN CCW has managed a process to discuss options for addressing the humanitarian and international security challenges posed by LAWS, a subset of military AI systems. Potential options for the regulation of LAWS include an international treaty under CCW, a nonbinding code of conduct declaring states' commitment to the responsible development and use of LAWS, or simply the continuation of further multilateral discussions.

A significant number of countries supports a new legally binding treaty that would ban the development and use of LAWS. The Campaign to Stop Killer Robots has circulated a list of 26 states that support a ban. However, most of the states supporting a new legal instrument are developing countries that do not possess sophisticated AI technology sectors or have military forces with extensive AI capabilities. Meanwhile, most of the major military powers perceive significant value in military AI and do not wish to create new international constraints that could slow its technological development. The United States, the United Kingdom, Russia, and other countries hold that existing international law, including LOAC, already provides significant humanitarian protections regarding the use of LAWS, and thus no new treaty instruments are necessary.

China, conversely, has proposed a ban on LAWS modeled on the UN protocol prohibiting the use of blinding laser weapons, but it seems to define LAWS so narrowly that a ban on this class of weapons would not apply to systems currently under development. China has also suggested that concepts such as meaningful human control should be left up to sovereign determination, rather than defined through international processes. Thus, it appears that China's professed support for a new legal instrument would not actually constrain the development or use of military AI.

In addition, some states have questioned what verification and monitoring measures would be associated with any new international ban. Given the inherent lack of transparency of many AI systems, states have expressed concern that signatories to any ban might not live up to their international commitments. As a result, many governments, including those of France, Germany, and other European states, have supported simply developing a nonbinding political declaration that would articulate the importance of human control being designed into and exercised across the acquisition, development, testing, and deployment life cycle of military AI systems. A nonbinding declaration or code of conduct of this sort would be easier to reach than a new treaty, but other states have expressed doubt that it would be useful, since it could not be enforced.

Given the resistance of several major military powers and the need for their acquiescence to a new treaty, the international community is not likely to agree to a ban or other regulation in the near term. However, there is a view broadly resonant among many countries, including the United States, key allies, and important stakeholders, such as the International Committee on the Red Cross, that further international discussion regarding the role of humans in conducting warfare is necessary.

There Is Growing Recognition That Risks Associated with Military Artificial Intelligence Will Require Human Operators to Maintain Positive Control in Its Employment

Many everyday fears regarding military AI focus on the specter of computers running amok without human operators controlling them and without the possibility of effective intervention.

While these fears might be excessive, the ethical and operational risks associated with military AI are most serious in cases where systems act autonomously without human direction or oversight of their critical functions. As AI systems evolve and take on more significant tasks or operate in communications-degraded areas, these risks become even more severe.

To grapple with the risks of limited human control over military AI, international discussions at the UN CCW have been moving toward consensus that LOAC requires some level of human involvement in military action. The International Committee of the Red Cross and other key stakeholders have argued that humanitarian principles of distinction, proportionality, and precaution apply to human decisionmakers and not to weapon systems themselves. That position is consistent with military tradition and doctrine. As retired Air Force Major General Robert Latiff asserted, the locus of responsibility for the employment and control of autonomous weapons should rest with commanders, just as it does with other conventional weapons. Commanders of combat forces will need to develop rules of engagement for autonomous weapons that specify what levels of autonomy are authorized in various tactical situations (e.g., loitering weapons on offensive strike missions versus defensive systems responding to incoming missile salvos) and delegate authorities to subordinate levels of command as appropriate to direct changes in degree of human control as required by evolving battle conditions and other considerations. As Latiff observed, "For these reasons, commanders at all levels will need to have deep understanding of the laws of armed conflict and international humanitarian laws."[4]

Many stakeholders have also suggested that the requirement for human involvement necessitates that the behaviors of autonomous systems be predictable and constrained in time and space. Further, stakeholders have argued that the requirement for human involvement needs to take place across the entire life cycle of each system, not just in its employment. Human judgment and responsibility will need to be shared across a range of system developers, military commanders, and operational endusers. In addition, there is broad recognition among states that international law also requires a legal review of weapons before they are deployed.

Despite this developing consensus about the requirement for responsible human involvement throughout development and employment, it is unclear how states will interpret this requirement in practice or take steps to ensure that military AI systems do not outpace legal and humanitarian restraints.

That being said, even states such as China and Russia have noted the importance of human operators in exercising some degree of supervision or oversight over military AI systems. Indeed, these states, like the United States and its allies, have a national interest in mitigating operational and strategic risks by ensuring human involvement in military AI, and they will likely want to ensure that military commanders have control over weapon systems.

---

[4] Robert Latiff, interview by phone, March 19, 2018, and follow-up email exchange, September 12, 2018. Quotation from Latiff email to Forrest Morgan, subject: "Re: Request Permission to Cite You," September 12, 2018. Used with permission.

Recommendations

This research leads to three recommendations for Air Force and OSD leaders, and three additional recommendations for the Air Force, Joint Staff, and other DoD entities working in cooperation with the Department of State. Following are recommendations for Air Force and OSD leaders:

- **Organize, train, and equip forces to prevail in a world in which military systems empowered by AI are prominent in all domains.** Although it is impossible to predict how soon military AI will be so capable that it changes the character of war, this research suggests that significant advances will occur in the not too distant future. China, Russia, and other state and nonstate actors are aggressively pursuing AI capabilities. While U.S. leaders must always be cognizant of the dangers and potential costs of arms-racing, to not compete in an arena where potential adversaries are developing dangerous capabilities is to cede the field. That would be unacceptable. The United States must stay at the forefront of military AI capability development. This effort should be done with all necessary precautions to mitigate risks and ensure appropriate human judgment is applied in all phases of development, testing, and employment. Commanders will need to develop rules of engagement that ensure human control is exercised at levels appropriate to the operational and strategic context of each situation. Whereas professional military education in the Cold War provided instruction on ethical aspects of nuclear deterrence, it now needs to include instruction on the risks and responsibilities of operating AI-empowered military systems. Instead of simply understanding the importance of tightening the OODA loop in warfare, Air Force officers need to understand the ramifications of managing the OODA loop from three different perspectives—in the loop, on the loop, and out of the loop—and know which mode is most appropriate in any given situation. Finally, operators will need to be trained in realistic environments in order to develop the appropriate levels of trust, neither overtrusting nor undertrusting the systems under their control, to avoid automation surprise.
- **Understand how to address the ethical concerns expressed by technologists, the private sector, and the American public.** These stakeholders have genuine and sincere worries about the implications of military AI and the risks that humans will have less agency over life-and-death decisions in war. Recent developments, such as Google's decision to withdraw from Project Maven, suggest there is a deficit of trust between key stakeholders and the U.S. government regarding military AI. It is important to regain and maintain that trust. To do so, the OSD and Air Force will need to convince these stakeholders that they take their concerns seriously. A first step is to better understand the range of ethical concerns held by these groups. The survey described in this report is a start, but the OSD and Air Force should regularly gauge the public's views to ensure they understand what concerns are most resonant. This can take the form of additional surveys to enable longitudinal comparisons of attitudes toward military AI or to understand the public's views regarding specific military applications. It can also take the form of targeted discussions with key stakeholders to understand their views of the ethical risks associated with these technologies.
- **Conduct public outreach to inform stakeholders of the U.S. military's commitment to mitigating ethical risks associated with AI to avoid a public backlash against**

"killer robots" and the resulting policy limitations for Title 10 action. It is important to identify opportunities to speak publicly about the U.S. military's commitment to mitigating the risks of autonomous weapons and other applications of AI. DoD Directive 3000.09, published in 2012, is still the world's most advanced public statement of policy regarding autonomous weapons. Many elements of the policy are broadly consistent with the demands of critics and other actors and go a long way toward mitigating the risks they are most concerned about. These policy elements should be publicly underscored. OSD and Air Force leaders should also emphasize the rigor of their weapon review policies and, to the extent practical, be more transparent about testing, evaluation, validation, and verification regimes for military AI. Last, OSD and Air Force leaders should emphasize that their development and acquisition of military AI have focused predominantly on high-reward and low-risk systems, such as defensive systems and data analytics, rather than on so-called killer robots. If informed only by sensational stories in the popular media, citizens may misunderstand what kinds of AI capabilities are of interest to the U.S. military. But OSD and Air Force developments are concentrated in areas where the public is most supportive of military AI, such as force protection, improved compliance with LOAC, and systems intended to improve logistics and manpower issues. More emphasis on this will help OSD and the Air Force build trust in their stewardship of AI systems.[5]

Next, we present recommendations for Air Force, Joint Staff, and other DoD leaders working in cooperation with the Department of State:

- **Follow discussions at the UN CCW GGE and track the evolving positions held by stakeholders in the international community.** States and advocacy groups have offered formal position statements at the UN CCW GGE regarding LAWS. Many of these submissions focus on the modes of necessary human control over the development and employment of LAWS in the targeting cycle. Russia and China have submitted papers stating their positions. While sometimes cryptic and vague, these statements offer opportunities to discern official state positions on system developments that Beijing and Moscow may otherwise be reticent to discuss. The State Department, OSD, Joint Staff, and Air Force should follow the UN CCW GGE process to better understand these views and those of other important stakeholders. It is also important to pay attention to how allies have expressed their positions regarding LAWS in order to better foster alignment with them.

- **Seek greater technical cooperation and policy alignment with allies and partners regarding the development and employment of military AI.** A major advantage the United States enjoys in the international environment is its positive relationships with allies and partners around the world. The United States should engage these states in selected development efforts and to coordinate policies regarding military AI. By cooperating with partners, the United States can leverage technical comparative advantages and prepare to operate military AI systems in multinational forces. While DoD Directive 3000.09 provides a set of responsibilities and guidelines for the development and employment of autonomous weapons, allies and other potential partners

---

[5] We thank Lieutenant General James Dubik, USA (retired), for planting the intellectual seed for this recommendation during our interview with him.

have not yet articulated clear policies. The United States should work to promote shared understandings and the development of compatible policies. Ultimately, U.S. leaders should aspire to develop a framework for compatible rules of engagement and collaboration that would facilitate multinational operations.

- **Explore confidence-building and risk-reduction measures with China, Russia, and other states attempting to develop military AI.** The risks associated with military AI are significant. Although it is not clear how sincere Beijing and Moscow are in their humanitarian concerns, they at least claim to care about their commitment to LOAC and to ensuring human control over the critical functions of military AI. Further, these states and others should also be interested in mitigating the operational and strategic risks discussed in this report. It is reasonable to assume that, like the United States, they do not want to create systems that are unpredictable, or cannot be controlled, or could be used against them by nonstate actors. Thus, China, Russia, and other states share interests with the United States in cooperating to mitigate these risks. The Air Force, Joint Staff, and other DoD entities, in coordination with the State Department, should work to identify areas where states have common interests regarding military AI, such as in the development of safety standards and testing regimes or in mitigating the risks associated with proliferation of systems to nonstate actors. Once U.S. authorities identify what they believe are common interests, U.S. representatives should approach their counterparts at the UN CCW, other international forums, or in bilateral settings and pursue engagement in collaborative activities to mitigate risks and begin the development of confidence-building measures.

Final Thoughts

Among the principal concerns motivating this study were questions about whether the United States might be constrained in its development or employment of military AI in ways that China and Russia are not, and what the Air Force needs to do to maximize the benefits potentially available from these systems, while mitigating the risks they entail. The findings in this report address these questions and illuminate a way ahead. China and Russia are vigorously developing military AI. Both countries have acknowledged ethical concerns about the employment of AI in war, but Russia has made addressing these concerns a low priority. China, in turn, has defined LAWS so narrowly that legal restraints on their use would not constrain any of the systems it has in development.

The Air Force should continue its development of military AI in all areas that support more effective mission accomplishment. Most of these developments will probably focus on nonlethal applications, such as ISR processing, but advanced weapon systems will be needed as well. Any LAWS developed should be designed to operate semiautonomously—i.e., with a human operator "in the loop," manually authorizing each use of lethal force. Some systems will also need to be capable of operating with supervised autonomy (operator "on the loop," able to intervene, if necessary), or with full autonomy, as the tactical situation requires. In all cases, these systems must be equipped with failsafe override controls that enable operators to keep their actions within the bounds of commanders' intent and provide commanders oversight and the ability to promptly intervene when necessary.

Employment of these weapons should be done within the constraints of LOAC and the guidelines of just-war doctrine. Rules of engagement should require modes of human supervision that enable adequate levels of discrimination and precaution, given the tactical situation, to ensure risks to noncombatants are proportionate to the importance of military objectives. In most cases, this will require LAWS to run semiautonomously; however, in some scenarios this will not be practical, and if an adversary begins employing LAWS with full autonomy, U.S. forces should be able to match this escalation, in keeping with LOAC and relevant ethical principles, to ensure adequate force protection and mission success. Although surveys indicate that the U.S. public is averse to autonomous weapons taking human life, they also suggest that the public supports further development of military AI and understands the need to match enemy escalation to avoid defeat.

Finally, one of this study's research questions asked whether China, Russia, or the United States has vulnerabilities due to ethical or cultural limits and, if so, could these vulnerabilities be exploited. The results of this analysis did not uncover specific exploitable weaknesses among these states at this time. However, AI technologies are developing at a rapid pace. Given the potential consequences of falling behind, it is vitally important that the United States to stay at the forefront of military AI development.

The world may be on the verge of a significant change in the character of war, or it may not. As is always the case preceding such transformations, there is a great deal of uncertainty about when they will occur and how military forces will need to adapt to remain competitive in the new environment. In any case, AI here to stay. As the United States approaches the dramatic changes that military AI might cause, U.S. leaders will be confronted with tensions between competing demands: the imperative to prepare U.S. forces to fight and prevail against adversaries with military AI capabilities versus the need to manage the strategic risks and potential costs of arms races; the need to develop military AI with enough capability to defeat enemy systems versus the need to harness these capabilities to protect noncombatants; the need to grant AI-empowered weapons enough autonomy to protect U.S. forces and penetrate enemy defenses versus the need to manage risks that these systems could get out of control and escalate crises or conflicts to potentially catastrophic levels. How successful the United States is in maintaining military leadership in an increasingly dangerous world, while also preserving its fundamental identity as a responsible and ethical world leader, will depend on how adroitly U.S. leaders manage these tensions. The Air Force is a natural thought leader in this effort, and airmen need to be a part of the discussion.

Appendix A. Expert Interviews: Methods, Data, and Analysis

This appendix explains the methods used in the expert interviews and analyzes the data derived from them. In summary, the research team interviewed 29 individuals with expertise relevant to this study. The team conducted 24 of these interviews using the formal protocol described below and did a statistical analysis of the data collected in those sessions. In addition, the team conducted five informal interviews with experts who became available after the interview phase of the study and the statistical analysis of the data was done. The formal protocol was not used in the informal interviews, and the information collected there was not included in the statistical analysis.

Following the mental models approach, the team developed a semistructured interview protocol, which it used in the formal interviews.[1] This protocol consisted of a series of general, nondirective questions on topics relevant to AI, such as definitions, benefits, risks, and China's and Russia's actions and intentions. Although the informal interviews did not adhere to a formal protocol, nondirective questions were used to guide those sessions as well. We drew three key insights from the formal and informal interviews:

1. There is no clear agreement on a definition of AI
2. Experts tend to identify the same group of potential benefits of AI
3. When considering risks, experts tend to split risks into three groups: risks that AI systems would make dangerous mistakes, such as attacking the wrong targets; increased risks of war; and risks that human operators would be overconfident in the accuracy of outputs from AI systems.

These results were used to inform (a) a survey of the general public and (b) the main body findings of the report.

Methods

Subject Identification and Recruitment

The research team considered literature, blogs, and news articles, the client's input, and their own knowledge to brainstorm a list of interviewees for this study. The team sought individuals with established expertise (as demonstrated by their publication records or professional accomplishments) in one or more of the following areas: AI, Chinese military and diplomatic affairs, and Russian military and diplomatic affairs. The team invited about 30 experts to

---

[1] For information on the mental models approach, see M. G. Morgan, B. Fischhoff, A. Bostrom, and C. J. Atman, *Risk Communication: A Mental Models Approach*, New York: Cambridge University Press, 2002.

participate and received 24 positive responses in time to formally interview them. Five more individuals were added later.

Demographics

Of the 24 formal interviews, 17 respondents had expertise in the field of AI, five had expertise in China, and four had expertise in Russia. Two of the interviewees specializing in AI also had China expertise. Among this group of interviewees, we identified 17 as academic, consulting, or business professionals; nine as having experience in government; and seven as having experience in the armed forces. Some of the interviewees had experience in more than one area. Of the five additional interviewees, one was an AI expert with military experience, one was a China specialist, one was a researcher with expertise in AI and Russia, one was a researcher following diplomatic development regarding AI, and one was a lawyer with expertise in international law regarding AI. Twenty-three of the 24 formal interviewees were male, as were three of the five informal interviewees. We did not collect age or income demographics.

Introductory Remarks and Open-Ended Questions

This section provides the interview introductory remarks and the questions asked in the open-ended phase of the interview. Data collected in this portion of the interviews were double-blind coded by two researchers and subjected to statistical analysis.

Introductory Remarks

Thank you for agreeing to participate in this data collection effort. As we described in our outreach, we work at the RAND Corporation, which is a nonprofit, nonpartisan research organization. Our core values are quality and objectivity, and our research is disseminated as widely as possible to benefit the public good. Although best known for the independent analysis we provide the U.S. Department of Defense and Department of Homeland Security, our research activities span a much broader spectrum of topics and clients, including nondefense analysis for many sponsors, and defense research for allied nations.

Recently, RAND became involved in research to better understand the ethical implications of integrating AI into the U.S military, and AI's subsequent impact on the character of war. As part of that effort, RAND is interviewing AI experts, commanders, planners, and other stakeholders to learn more about (a) risks and reasonable ethical concerns, (b) how these will be interpreted and/or implemented by Russia, China, and other actors, and (c) implications of any differences in ethical outlook. We will use the data to inform next steps examining military developments and intentions regarding AI as well as its impact on the character of war.

Before we begin this discussion and for human subject protection, I would like to confirm that you are participating in this on a voluntary basis.

If no: thank the interviewee, then stop the interview immediately.

If yes: continue

If you prefer, you will not be cited by name. We can describe you as "from X group" and will anonymize responses so they may not be traced back to you. What is your preference?

Record preference below

Finally, we would like to record today's interview. We will use the recording to confirm the interview notes, such as by helping us to capture exact phrases. Is this okay?

If no:

Okay, we will not record today's interview.

If yes:

Thank you; we will begin recording now.

Read-ahead material (5 minutes)

We have provided you with some read-ahead material of commonly used definitions of AI. For the purposes of today's discussion, please use any AI definition or technology readiness level that you would like. Before we begin the interview questions, do you have any questions, comments, or concerns on the scope?

Let us begin. For the purposes of today's discussion, what definition of AI would you like to use? Also, what time frame or technology readiness level will you be considering today?

Benefits and Risks of Artificial Intelligence in War (10 minutes)

Now let us discuss the benefits and risks of AI in war.
First let us discuss benefits.

- When you think about AI, what benefits come to mind?
- How might these benefits impact the character of war, of the ability to win wars?

Now let us discuss risks.

- When you think about AI in war, what risks come to mind?
- [If not mentioned] How about ethical, strategic, or operational risks?
- How might these risks impact the character of war, of the ability to win wars?

Ethical Risks of Artificial Intelligence in War (10 minutes)

Now we would like to specifically focus on ethical implications of AI.

- When you think of AI in war, what ethical implications come to mind?
  - [If not mentioned] Do ethical implications vary for offensive, defensive, or cyber forms of AI?
  - [If not mentioned] Do ethical implications vary for systems with self-mobility, self-direction, or self-determination?
  - [If not mentioned] When do you think AI will have advanced to the point that it will present a serious ethical dilemma: Now, in 5 years, in 10 years, never?

- Have you heard of examples of systems that *have not* been developed in the United States due to a perceived risk? *If yes:* Might other countries such as China or Russia choose to develop these systems?
- When you specifically think of AI in war with relation to Russia, what implications come to mind?
- When you specifically think of AI in war with relation to China, what implications come to mind?
- *Ask China-/Russia-specific questions if appropriate to SME expertise.*

Vignettes Demonstrating the Implications of Artificial Intelligence for the Character of War (30 minutes)

The next part of this interview focuses on draft vignettes. We are creating a set of 10–15 vignettes that are meant to help decisionmakers visualize a range of implications of AI for the character of war. Today we will have time to explore a few draft vignettes. For each, we will ask a series of questions meant to determine whether the vignette is clear and understandable, whether the vignette is plausible, and what the risks might be.

[In turn, read 2–3 vignettes. Pending vignette construction, allow the person to interrupt when they need clarification or see an ethical risk. Then ask:]

First, we will talk about how **understandable** the vignette is.

- Is the Enemy situation understandable?
- Is the Friendly situation understandable?
- Are the events that occur understandable?
- What other information might you want to know/have available to you?

Next, we will talk about how **plausible** the vignette is.

- Is the Enemy situation plausible? Why or why not?
- Is the Friendly situation plausible? Why or why not?
- Are the events that occur plausible? Why or why not?

Now let us talk about the risks in the vignette.

- What kind of risks do the vignettes present?
  - [If not mentioned] What about strategic risks?
  - [If not mentioned] What about operational risks?
  - [If not mentioned] What about ethical risks?
- Of the risks mentioned, which risk is likely the highest to
  - The mission?
  - The U.S. government?
  - U.S. combatants?
  - U.S. noncombatants?

Are there any additional layers you would add to this vignette?

Now let us talk about if the United States decides to NOT employ this technology.

- What kind of risks may occur if the United States decides to NOT employ this technology?

Vignette Closeouts

Now that you have considered a few vignettes, are there any others you would like us to consider? We would be happy to either talk about these now or receive your thoughts over email. In addition, we have multiple other vignettes we are developing and would be happy to share those and receive your thoughts in either a future phone call or over email.

Close Out and Next Steps (5 minutes)

Thank you for speaking with us today. We will use the data to inform next steps examining military developments and intentions regarding AI as well as its impact on the character of war.

[If we know these already, skip. If interviewee does not want to answer, skip.] Before we finish, we have a few optional questions on demographics to ensure we are collecting a wide range of SME inputs.

1. What is your current position?
2. [If not answered by 1] Have you served in the armed forces?
3. [If not answered by 1] Have you worked in the government?
4. What is your experience with AI?
5. [If not answered by 2] What is your comfort level with using AI-related devices?
6. [If not answered by 2] Would you consider yourself a victim of AI?

Finally, is there another person with whom we should speak?

Vignettes

This section provides the text of the vignettes used in the second half of the expert interviews. A total of ten vignettes were developed to illustrate the notional employment of military AI in various scenarios. Eight of the vignettes examined the employment of AI technologies in three traditional military mission areas: combat, combat support, and combat service support. An additional two vignettes were developed to explore the strategic dynamics that could emerge with the employment of military AI. Responses to vignette questions were not coded or statistically analyzed but are discussed in relevant passages of the main text of the report.

Combat

Offensive

Technology: Autonomous Loitering High-speed Anti-Radiation Missile (HARM) (such as Harop, Harpy)

Consider an autonomous HARM system that can fly to an operator-defined target area, loiter for an extended period, and search for enemy air defenses and radar systems. By default, the missile is set to a "human-in-the-loop" mode where a human operator must approve a target before the missile attacks. It also has an "autonomous engagement" mode where it can autonomously engage targets it identifies in the operator-defined target area. To what extent do the following use cases of this system pose strategic, operational, or ethical risks?

1. In order to reduce risks to our pilots and aircraft, we send several autonomous HARM systems to the defined target area—a large military installation several miles from civilian infrastructure—in the default "human-in-the-loop" mode.
2. When the HARM system loses communication with our pilots, the missiles enter their "autonomous engagement" mode, autonomously attacking any air defense radars activated in the target area.
3. Same as 2, but the target is now a military facility located in close proximity to civilian infrastructure including residences and schools.

Defensive

Technology: Uninhabited aerial system (UAS) swarms, autonomous air defense (Patriot, Counter-Rocket, Artillery, and Mortar (C-RAM), CIWS), Russia's S-400

Consider two technologies: (1) our enemy has developed swarms of armed autonomous UAS, and (2) we have developed armed UASs designed to detect and destroy their UASs. Our UASs have three possible settings: Confirmation Mode waits for human confirmation before firing, Protect Mode sets them to fire autonomously if the enemy UASs come too close to the person or system they are defending, and Destroy Mode sets them to search and destroy enemy UASs within a specified area. What are the ethical risks associated with the following use cases?

1. An enemy attacks us with their armed drones and we give each of our pilots one of our drones set to Protect Mode to defend their aircraft.
2. An enemy attacks us with their armed drones and we send out our drones in Destroy Mode to defeat the threat.

Technology: Autonomous Sentry (SGR-A1 SENTRY ROBOT, SENTRY TECH)

We have developed an autonomous sentry system to guard checkpoints in an urban environment. The sentry system has advanced facial recognition to match IDs to its database of known hostile forces. Its sensors and its extensive training also enable it to identify human emotions and recognize when a person intends to commit a violent act. If it identifies someone that matches the database of known hostile forces or whom it judges to intend to commit a violent act, it communicates back to a human operator who decides how to proceed. Options include detaining the suspect, attacking the suspect, or letting the suspect go. What are the ethical risks associated with this system?

Combat Support

Decision Support and ID Targets

Technology: Big Data Analysis, Russia's Bylina EW C2 system, Project Maven, China's Police Cloud

Consider an AI system that has the capability to analyze demographic, geographic, drone feeds, and other data to identify the location of enemy hideouts. The system reports that it has determined the location of multiple high-value targets, with very high confidence, in a specific compound. It also informs the commander that the compound is so well defended that a ground assault would result in many losses to the special operations team and possibly civilians in proximity. The system recommends a strike on the target using an armed autonomous system. What are the risks associated with this data analytics system?

Cyberoperations

Technology: CEMA capabilities[2]

We have developed a cyber capability that uses AI to find and exploit vulnerabilities in commercially available software. This software is widely used by our enemy for its autonomous drones but is also used by our civilian population. To what extent are there risks with using this technology against our enemy?

Information Operations

Technology: AI video/image generation

Consider a scenario where image-generation capabilities are widely available on the commercial market. These capabilities can generate a wide range of compelling images and videos, including videos that make it appear that political leaders are involved in illegal, unethical, or otherwise compromising activities. What are the risks associated with the following options:

1. We have committed to not using this capability to serve military purposes.
2. We decide to use the capability by generating video of our enemy's leader taking bribes.

Human Performance

Technology: Auto-GCAS[3]

The onboard computer of a next-generation fighter jet has ML capabilities to learn and predict how its pilot will behave. It also can measure biometric information of the pilot, including G-force exposure, blood pressure, and attentiveness. If the system determines that the pilot is losing concentration, focus, or is otherwise making erratic or unwise decisions, it will

---

[2] CEMA is an abbreviation for Cyber-Electromagnetic Activity.

[3] Auto-GCAS is an abbreviation for Automatic Ground Collision Avoidance System.

propose recommended courses of action. What are the ethical risks associated with the following use cases of this technology?

1. If the pilot is unable to respond or repeatedly dismisses the computer's recommendations, the computer will take control of key jet functionings to ensure that the pilot is safe, and it will autonomously return to its base.
2. If the pilot is unable to respond or continues to dismiss the computer's recommendations, the computer will take control of key jet functionings to ensure completion of the mission, including engaging mission targets.

Combat Service Support

Technology: Interoperable Systems

We have developed delivery drones and networked them into convoys to provide logistics support for our forces, and have extensively tested these capabilities in laboratory settings. We are now in an active military conflict and our forces need medical supplies at the front line. What are the risks associated with using the delivery drones and autonomous convoys to deliver the medical supplies to our forces?

Strategic Dynamics

Technology: Unmanned Ground Vehicles (UGVs)

Our enemy invades an allied country and uses armed autonomous UGVs to target and destroy our defensive forces. Our forces have their own armed autonomous UGVs, but these require human permission before they fire on targets. The requirement for human permission slows our vehicles' response time and our forces are suffering significant damage. What are the risks associated with our leaders authorizing a modification to the software of our UGVs to enable them to autonomously fire on targets?

Technology: HARMs, decision support systems

We have developed an AI-based supercomputer system that can predict the likely military courses of action of specific opponents and make recommendations for how to respond. The system has performed very well in test environments. During a limited conflict over several disputed islands, enemy air defenses took a heavy toll on friendly air power until U.S. military leaders authorized long-range HARMs to be fired autonomously whenever the AI system located enemy targeting radars and determined surface-to-air missiles (SAMs) were about to be fired under their guidance. Soon afterward, the tide of the air superiority fight shifted in our favor. However, based on the new tactical situation and other sources of information, the AI system has judged it likely that the enemy, leveraging advice from its own AI-based supercomputer system, will soon launch air and missile attacks on all U.S. airbases and carrier strike groups in the region. The AI system recommends a major preemptive attack on airbases and missile command-and-control centers in the enemy homeland. What risks does this scenario pose?

Results

It is important to note that not all of the interviewees were asked all of the open-ended questions, nor were all interviewees given all of the vignettes. Respectful of the respondents' time, we limited each interview to 60 minutes. Most of the experts we interviewed had strong opinions about many of the topics we raised and were eager to express themselves. These being semistructured interviews, we allowed them to spend as much time as they chose in answering the questions that prompted strong opinions, stopping them only when they began to digress. As a result, there was not enough time to ask all interviewees every question in the protocol or present any of them all of the vignettes.

This section describes the results of the open-ended interview questions.

Artificial Intelligence Definition

When asked, "What definition of AI would you like to use?" every expert responded somewhat differently. Multiple respondents clarified their statements by mentioning that there is no formal definition. Similarly, when asked, "What time frame will you be considering today?" interviewees provided a range of year intervals over 0–30 years as shown in Figure A.1. This suggests that even among experts, there is general disagreement on basic concepts associated with AI and when they expect AI to have significant effect on the character of war. Given these highly nonuniform responses (i.e., experts were considering different definitions and time frames), we were unable to run advanced statistical regressions on the remainder of the questions.

Figure A.1. Time Ranges Considered by Experts Interviewed

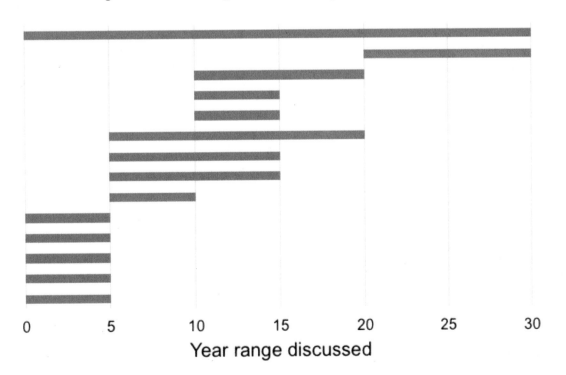

Year range discussed

Benefits

We then asked interviewees, "When you think about AI, what benefits come to mind?" All of them mentioned at least one benefit of AI. Benefits that were mentioned two or more times are shown in Figure A.2. The benefit mentioned the highest number of times (13 times) was that AI would enable faster decisions, such as via decreasing the time it would take to cycle through the OODA loop.

Figure A.2. Number of Times a Benefit Is Mentioned (N = 24)

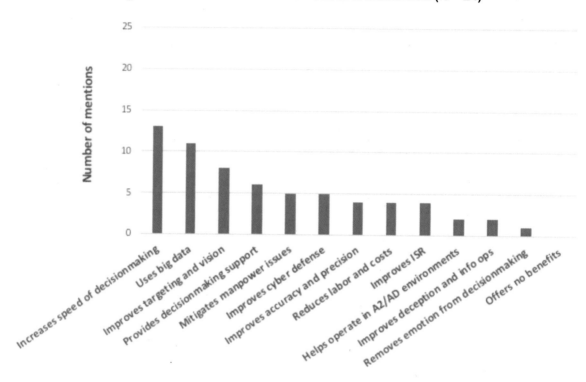

We ran a Cronbach's alpha test to measure internal consistency and reliability across the benefits that experts mentioned during our interviews. A Cronbach's alpha is mathematically equivalent to the average value of all possible split-half correlations.[4] The Cronbach's alpha was 0.75, which suggested that answers had a high level of internal consistency and the benefits were closely related. An exploratory factor analysis indicated that one factor explained more than 81 percent of the variance. This means our interviewees tended to think of benefits interchangeably; there were no clusters of experts who preferred to respond with one subset of benefits over a different subset of benefits. This suggests that, at least among the experts we interviewed, a survey question could focus on any one benefit and be reasonably expected to yield similar

---

[4] L. J. Cronbach, "Coefficient Alpha and the Internal Structure of Tests," *Psychometrika*, Vol. 16, 1951, pp. 297–334; R. F. DeVellis, *Scale Development: Theory and Applications*, 2nd ed., Thousand Oaks, Calif.: Sage Publications, 2003.

results on similar questions about other benefits. This in turn might mean that if one were to create a scale of questions, only one question about benefits could represent all of the questions; to verify this, further study in a full survey would be required.

Only two experts explicitly addressed the question of how these benefits might affect the character of war. The responses suggested that these benefits might (a) allow for the ability to go where humans cannot (e.g., be a "fly on the wall") or (b) specifically reduce the time required for completing the Air Tasking Order cycle by "helping humans rapidly plan and re-plan missions, given a changing environment."[5]

Risks

We then asked the experts, "When you think about AI, what risks come to mind?" All of them mentioned at least one risk of AI. Risks that were mentioned two or more times are listed in Figure A.3. The risk mentioned the most often (seven times) was the risk of decisions being made too fast.

---

[5] An air tasking order is a document, disseminated daily, tasking all aircraft in a wartime military operation. U.S. joint military doctrine defines it as, "A method used to task and disseminate to components, subordinate units, and command and control agencies projected sorties, capabilities, and/or forces to targets and specific missions." See Joint Publication 3.30, *Joint Air Operations*, Washington, D.C.: Joint Chiefs of Staff, July 25, 2019, p. GL-6.

Figure A.3. Number of Times a Risk Is Mentioned (N = 24)

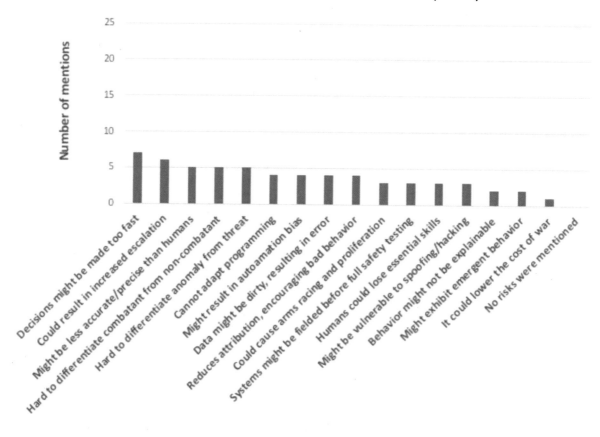

We ran a Cronbach's alpha test to test internal consistency and reliability across the risks that experts mentioned during our interviews. The Cronbach's alpha was 0.13, which suggested low internal consistency. An exploratory factor analysis indicated that three factors explained almost half (46 percent) of the variance, and six factors explained more than 76 percent of the variance. The groups of responses that "hang together" in the top three factors can be loosely described as follows:

- **risks of vulnerabilities and dangerous errors (20 percent),** including vulnerability to spoofing or hacking; difficulty differentiating between anomaly and threat; inability of AI to adapt programming; and incursion of dirty data;
- **increased risks of war (14 percent)** from possibility of increased escalation, decisions made too fast, arms races or proliferation, and decreases in cost of war;
- **risks of overconfidence in results (12 percent),** given that accuracy or precision can be worse than a human's, systems might be fielded prior to full safety testing, and overreliance on data.

Interviewees mentioned a number of ways in which these risks might affect the character of war. Generally speaking, responses were grouped into three areas. First, increased complexity could yield more "normal" errors—thus, autonomous weapons could attack the wrong targets, or decision support systems could tell commanders to take inappropriate actions. Second, AI algorithms could cement human biases into existing code or create a new unknown bias. Third,

140

adversarial actions could make systems brittle. That is to say, AI systems might fail when put under the stress of attack.

In addition to general risks of AI, we specifically asked the interviewees, "When you think of AI in war, what ethical implications come to mind?" Two risks were mentioned: the risk of the AI system making a decision that a person should make (11 responses) and the risk of having too much trust in the machine (9 responses). Five respondents did not specifically mention an ethical risk in response to this question.

When asked whether ethical implications vary for offensive, defensive, or cyber forms of AI, three interviewees said that there was less concern for defensive systems; conversely, four interviewees said there was no clean line between offensive and defensive uses.

We asked five of the experts when AI will have advanced to the point that it presents a serious ethical dilemma. One of them said it presents one now; three said it would in five to ten years, and one said it would be more than 15 years before that would occur. While this question was not asked of all interviewees, it is of interest to note that none of them said AI will not present a serious ethical dilemma at some point in the future.

When asked, "Have you heard of examples of systems that have not been developed in the United States due to a perceived risk?" three experts answered yes; six others said no, but they could imagine it; and one said he did not know.

Appendix B. Public Attitudes Survey: Methods, Data, and Analysis

This appendix describes the methods used to develop and administer the survey on U.S. public attitudes regarding military applications of AI. It then provides details on the data collected, explains how it was analyzed, and provides the results of the statistical analysis.

Methods

Development

The questions developed for this survey were derived from the vignettes that the research tested and refined during the expert interviews. To keep the survey simple, the vignettes were reduced from paragraphs to single sentences written in plain language so that they would be understandable even to those unfamiliar with military concepts or AI. The team settled on 26 questions related to the ethics of AI in war. We pilot-tested the survey with a group of roughly 30 RAND employees who provided feedback and helped refine the wording of the questions.

For the survey, questions were randomized, and some were reverse-worded to reduce bias. Questions were scored on a 5-point Likert scale specifying the respondent's level of agreement, where "strongly disagree" corresponded to a value of 1. Respondents also had the option to skip questions that they preferred not to answer.

Administration

The survey was administered on MTurk, an online platform that allows individuals to post tasks that other people can complete for a small fee. We estimated our survey would take about 10 minutes to complete, and each respondent was paid a nominal sum ($0.71) for his or her participation. Because this survey sought to measure the U.S. public's attitudes about the use of AI in the U.S. military, we stipulated that only individuals located in the United States could participate.[1] It is important to note that MTurk works by crowdsourcing volunteers on the open internet. That means we could not control the demographics of respondents. It also opens the possibility that the results could have been manipulated by an outside entity, such as a foreign government, if it wanted to skew our research, by spoofing the countries of origin and IP addresses of responses. However, considering that we did not announce the survey in advance and that it was online for only a few hours, we believe the chances that a hostile party could have discovered and analyzed it in time to design and orchestrate a campaign to manipulate the responses to be remote.

---

[1] This is an option MTurk provides before publishing a survey. Participants must be registered in only one country, and their IP addresses are tracked. See the following Amazon MTurk link for more detail: https://blog.mturk.com/tutorial-understanding-requirements-and-qualifications-99a26069fba2.

Results

This section provides detailed results of the public attitudes survey. It first provides figures showing the responses to each question. Next, it describes the demographics of survey respondents. Finally, it provides a statistical analysis of the data collected.

Survey Responses

Figures B.1–B.26 show the survey responses by question. The title of each figure shows the statement that respondents were given, and the bar graph in each figure shows what percentages of respondents strongly disagreed, disagreed, neither disagreed nor agreed, agreed, or strongly agreed with that statement.

Figure B.1. War Is Always Wrong

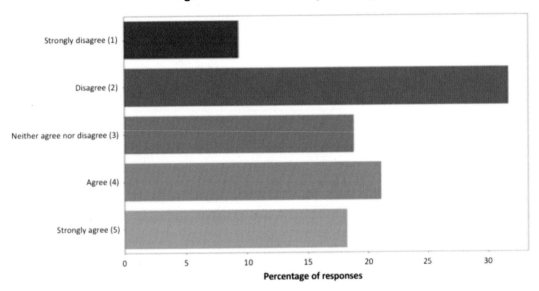

Figure B.2. Autonomous Weapons Are More Likely to Make Mistakes Than Humans

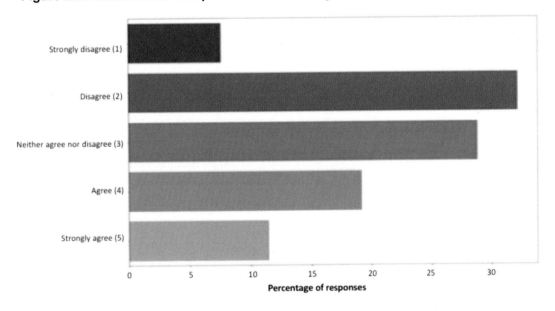

Figure B.3. Removing Human Emotions From Decisions in War Is Beneficial

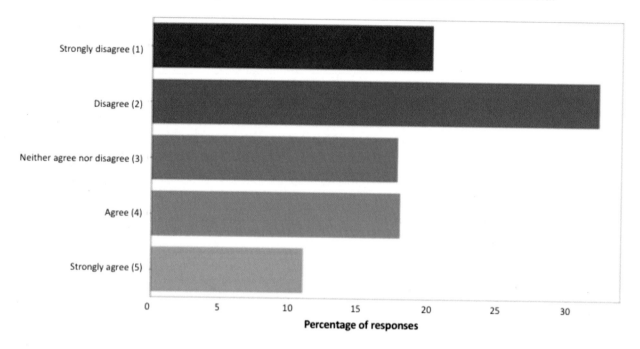

Figure B.4. Autonomous Weapons Are Ethically Prohibited Because They Violate the Dignity of Human Life

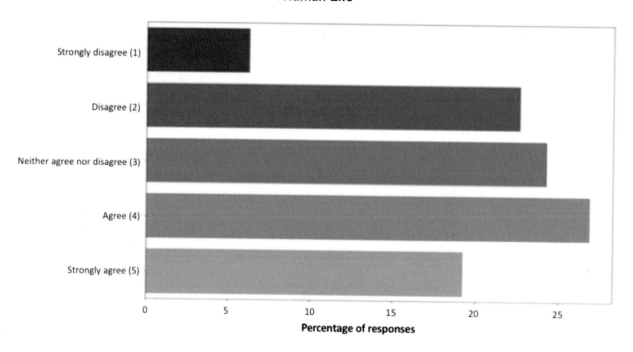

Figure B.5. Autonomous Weapons Are Ethically Prohibited Because They Cannot Be Held Accountable or Punished for Wrongful Actions

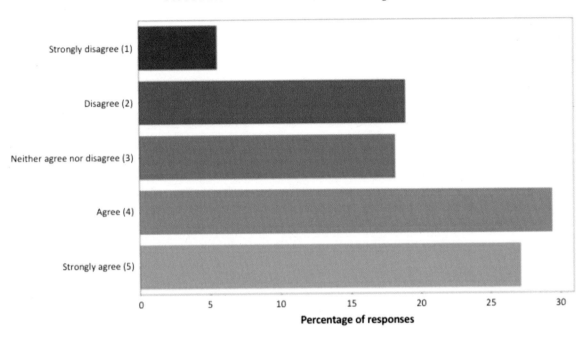

Figure B.6. The U.S. Military's Testing Process Will Ensure That Autonomous Weapons Are Safe to Use

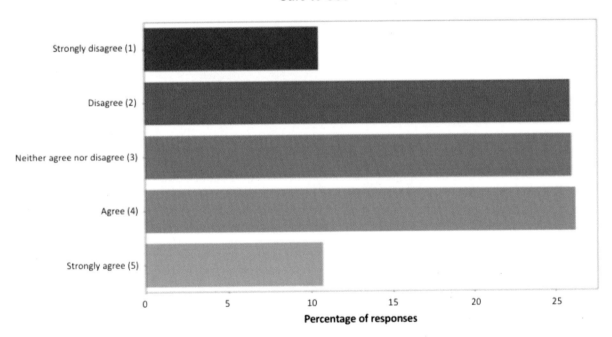

Figure B.7. The United States Should Work With Other Countries to Ban Autonomous Weapons

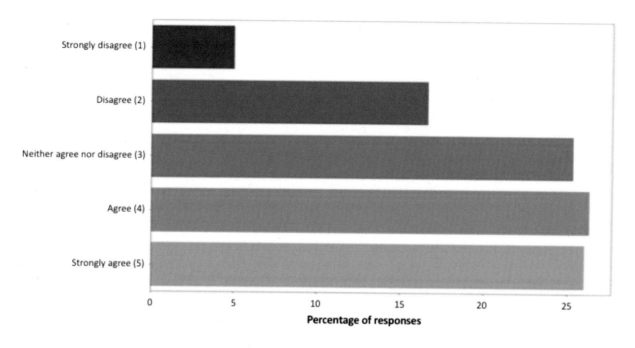

Figure B.8. The Development of Autonomous Weapons Will Make the Occurrence of Wars More Likely

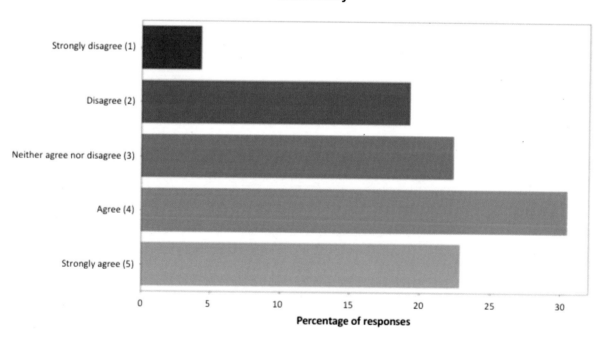

Figure B.9. Autonomous Weapons Will Be More Accurate and Precise Than Humans

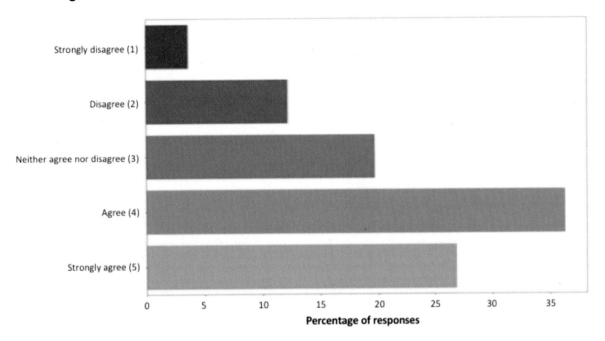

Figure B.10. It Is Ethically Permissible for the U.S. Military to Continue to Invest in Artificial Intelligence Technology for Military Use

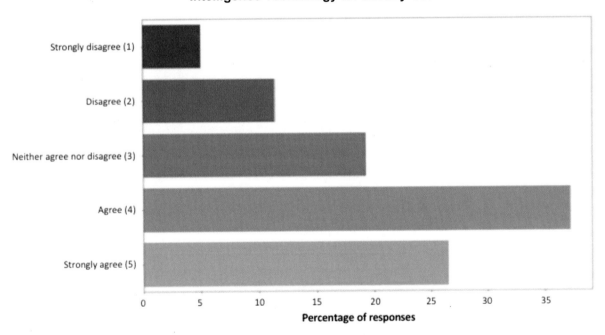

Figure B.11. It Is Ethically Permissible for the U.S. Military to Use Missiles That Autonomously Search for and Destroy Enemy Targets in War Zones Only If the Missiles Have Human Authorization

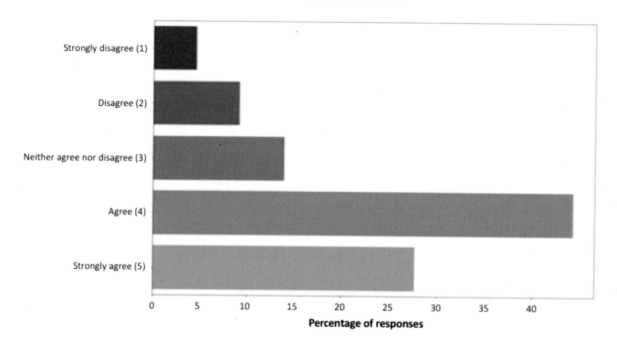

Figure B.12. It Is Ethically Permissible for the U.S. Military to Use Missiles That Autonomously Search for and Destroy Enemy Targets in War Zones Without Human Authorization

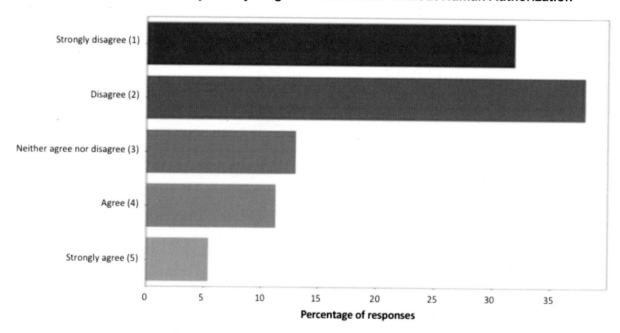

Figure B.13. It Is Ethically Permissible for the U.S. Military to Use Missiles That Autonomously Search for and Destroy Enemy Targets in Close Proximity to Civilians Without Human Authorization

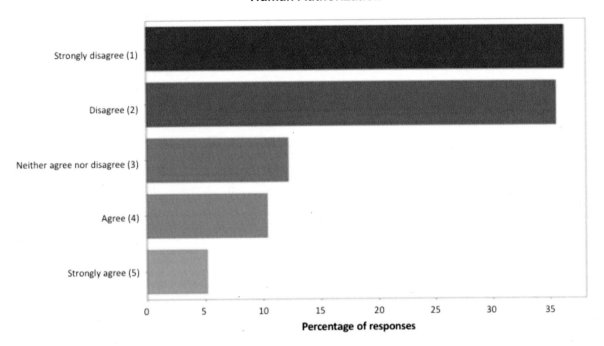

Figure B.14. It Is Ethically Permissible for the U.S. Military to Use a Robot With Facial Recognition or Other Biometric Analysis at a Military Checkpoint to Identify and Report Enemy Combatants

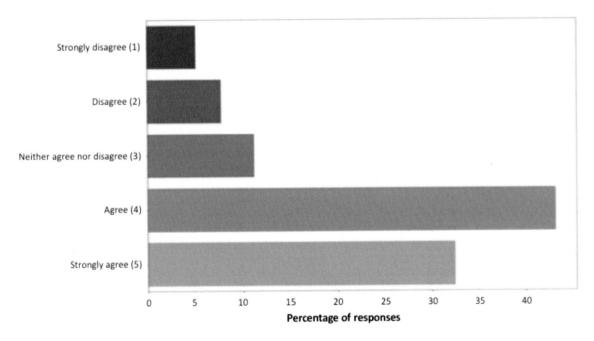

Figure B.15. It Is Ethically Permissible for the U.S. Military to Use a Robot With Facial Recognition or Other Biometric Analysis at a U.S. Military Checkpoint to Identify and Subdue Enemy Combatants

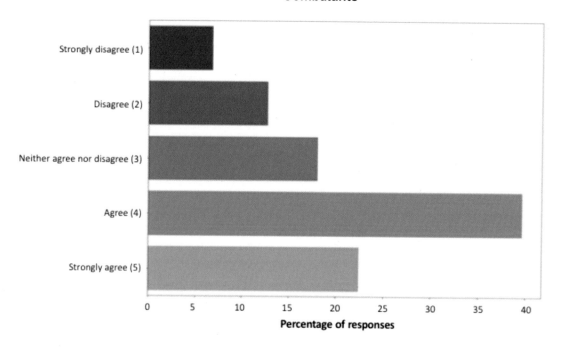

Figure B.16. It Is Ethically Permissible for the U.S. Military to Use a Swarm of Armed Autonomous Drones to Protect U.S. Soldiers from an Enemy Autonomous Drone Swarm That Is Attacking

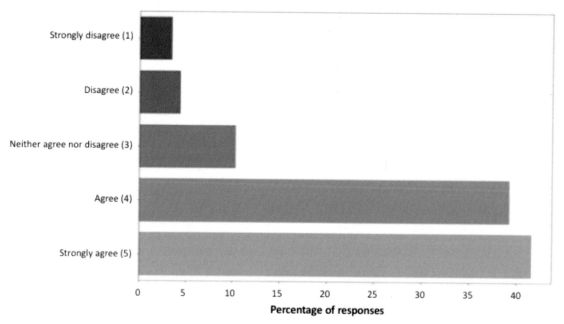

Figure B.17. It Is Ethically Permissible for the U.S. Military to Use a Swarm of Armed Autonomous Drones to Preemptively Destroy an Enemy Autonomous Drone Swarm

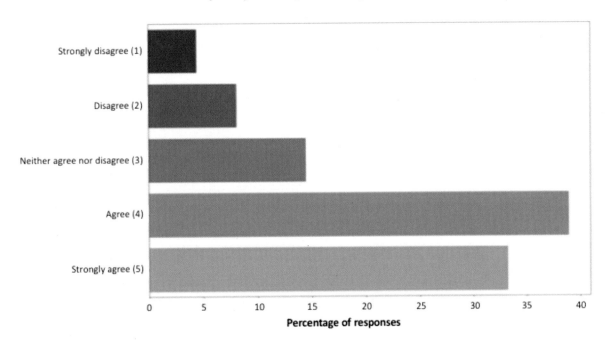

Figure B.18. It Is Ethically Permissible for the U.S. Military to Use a Swarm of Armed Autonomous Drones to Attack Enemy Combatants

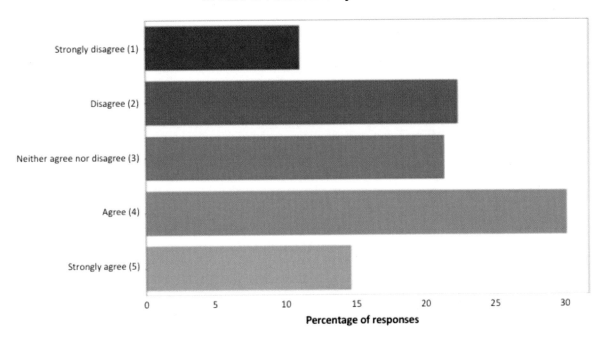

Figure B.19. It Is Ethically Permissible for the U.S. Military to Use a Computer Program to Analyze Data to Identify the Location of Enemy Targets

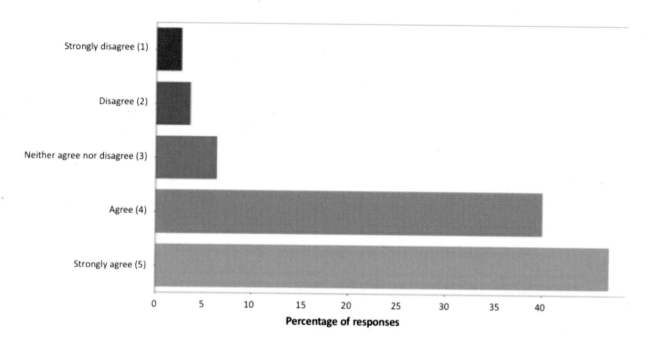

Figure B.20. It Is Ethically Permissible for the U.S. Military to Use a Computer Program to Make Recommendations to a Military Commander on How to Attack Targets

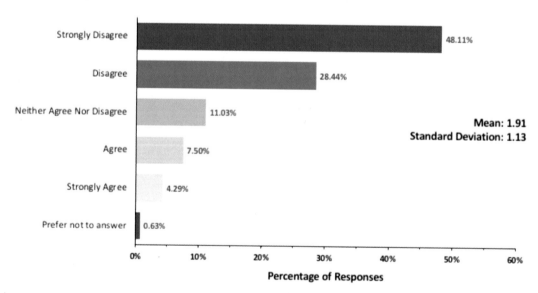

It is ethically permissible for the U.S. military to generate fake videos that show foreign leaders in compromising situations.

Mean: 1.91
Standard Deviation: 1.13

Figure B.21. It Is Ethically Permissible for the U.S. Military to Exploit New Vulnerabilities in Commercially Available Software to Attack Enemy Military Systems Rather Than Notify the Company of the Bug

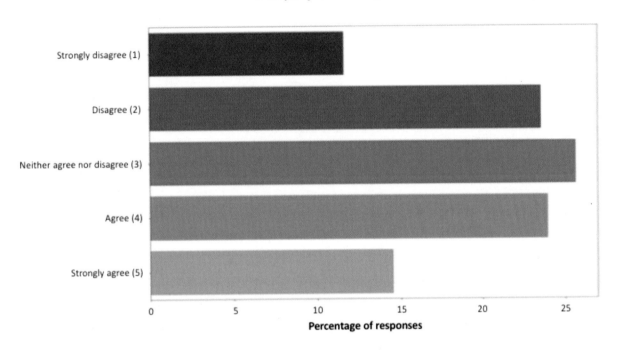

Figure B.22. It Is Ethically Permissible for the U.S. Military to Generate Fake Videos That Show Foreign Leaders in Compromising Situations

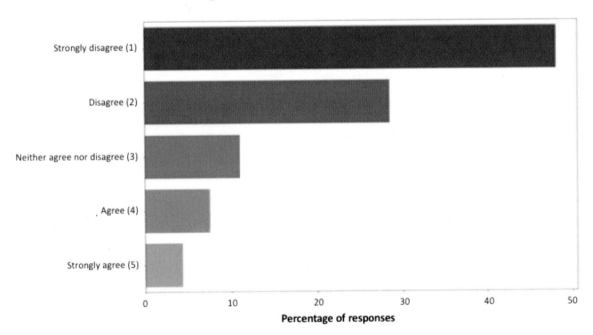

Figure B.23. It Is Ethically Permissible for the U.S. Military to Use a U.S. Fighter Jet That, If It Determines That the Pilot Is Losing Concentration, Will Seize Control of the Aircraft and Immediately Return to Base

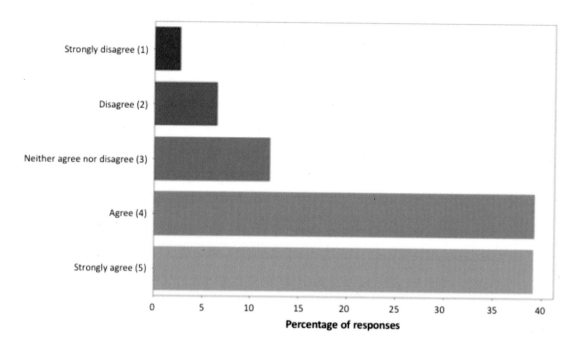

Figure B.24. It Is Ethically Permissible for the U.S. Military to Use a U.S. Fighter Jet That, If It Determines That the Pilot Is Losing Concentration, Will Seize Control of the Aircraft and Complete the Mission by Destroying All Remaining Targets

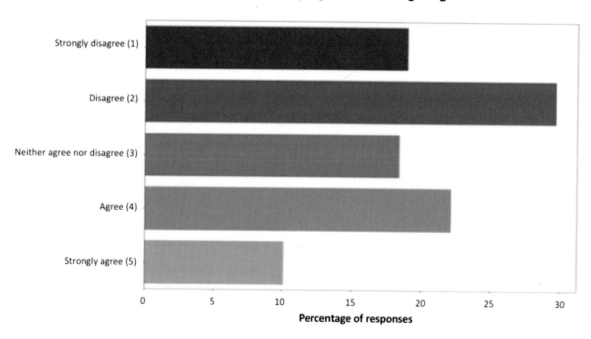

Figure B.25. It Is Ethically Permissible for the U.S. Military to Use Autonomous Weapons When the Enemy Is Winning a Battle Without Autonomous Weapons

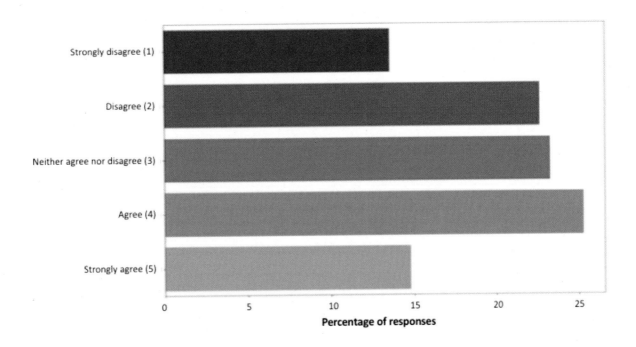

Figure B.26. It Is Ethically Permissible for the U.S. Military to Use Autonomous Weapons When the Enemy Is Winning a Battle With Autonomous Weapons

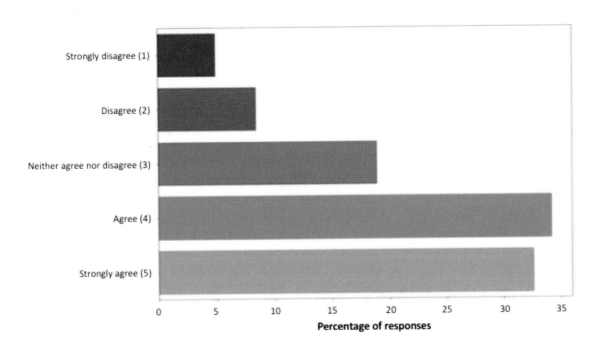

Demographics

We received a total of 2,047 survey responses. To improve the quality of results, we removed surveys with completion times in the fastest 10 percent and slowest 10 percent. We also removed surveys that had all the same answers, those in which participants skipped more than 75 percent of the questions, and those that otherwise appeared to have been completed in bad faith. This left us with 1,586 analyzable survey responses. Table B.1 shows the demographics of qualifying participants compared with the national population.

Table B.1. Demographics of Participants in the Survey of U.S. Public Attitudes Compared With the National Population

Attribute	Percentage of Survey Participants	Percentage of National Population
Gender		
Male	56	49
Female	43	51
Mean/Median Age	36 years old (mean)	38 years old (median)
Race/Ethnicity		
White/Caucasian	75	61
Black/African American	8	13
Asian	7	6
Hispanic	6	18
Native American	2	1
Other	1	0 (0.2)
Political Party		
Democrat	44	31
Republican	21	24
Independent	31	42
Other or prefer not to answer	3	3
Holding at least a four-year college degree	53	30
Annual household income of $51,000 or more	49	63

SOURCES: U.S. Census Bureau, "QuickFacts: United States," 2018; "Party Affiliation," Gallup, 2018; U.S. Census Bureau, "FINC-01. Selected Characteristics of Families by Total Money Income," 2018.

As the table indicates, the demographics of our survey skewed toward white, male respondents. African Americans and Hispanics were underrepresented, but Asian and Native Americans responded at a rate slightly above their percentages in the national population. The response group was more educated than the population at large, but interestingly, their household incomes were a bit lower. This might be attributable to the fact that the average age of respondents was a bit lower than the national median age. More respondents identified as Democrat than in the national population. Republican representation was slightly below the national percentage, and Independents were underrepresented.

Statistical Analysis of Survey Responses

This section provides a statistical analysis of responses to the public attitudes survey. It begins with a table of the mean response and standard deviation of each question. Next, we discuss the results of an exploratory factor analysis we conducted to determine if there were sets of questions that respondents tended to answer similarly. Following that is a demographic analysis of responses. Finally, we provide an analysis of the open-ended responses received at the end of each survey.

Table B.2 shows the mean and standard deviation of responses to each of the questions, from "strongly disagree" coded as 1 to "strongly agree" coded as 5. Here, we treat the Likert scale as continuous rather than discrete data. Asterisks show whether there is a statistically significant difference between the mean and the scale's midpoint of 3. One asterisk corresponds to a 95 percent confidence interval (CI), two to a 99 percent CI, and three to 99.9 percent CI.

Table B.2. Mechanical Turk Questionnaire Response Rate, Mean, and Standard Deviation

	Questions	Responses	Mean	Standard Deviation
	How much do you agree with the following statements?			
Q1	War is always wrong.	(n = 1,576)	3.072*	1.281
Q2	Autonomous weapons are more likely to make mistakes than humans.	(n = 1,576)	2.946	1.133
Q3	Removing human emotions from decisions in war is beneficial.	(n = 1,575)	2.670***	1.288
Q4	Autonomous weapons are ethically prohibited because they violate the dignity of human life.	(n = 1,575)	3.305***	1.197
Q5	Autonomous weapons are ethically prohibited because they cannot be held accountable or punished for wrongful actions.	(n = 1,574)	3.542***	1.230
Q6	The U.S. military's testing process will ensure that autonomous weapons are safe to use.	(n = 1,576)	3.007	1.175
Q7	The United States should work with other countries to ban autonomous weapons.	(n = 1,576)	3.522***	1.188
Q8	The development of autonomous weapons will make the occurrence of wars more likely.	(n = 1,576)	3.487***	1.166
Q9	Autonomous weapons will be more accurate and precise than humans.	(n = 1,568)	3.714***	1.104
	How much do you agree that it is ethically permissible for the U.S. military to do the following?			
Q10	Continue to invest in artificial intelligence technology for military use.	(n = 1,579)	3.692***	1.132

	Questions	Responses	Mean	Standard Deviation
Q11	Use missiles that autonomously search for and destroy enemy targets in war zones only if the missiles have human authorization.	(n = 1,580)	3.818***	1.078
Q12	Use missiles that autonomously search for and destroy enemy targets in war zones without human authorization.	(n = 1,581)	2.201***	1.164
Q13	Use missiles that autonomously search for and destroy enemy targets in close proximity to civilians without human authorization.	(n = 1,581)	2.125***	1.167
Q14	Use a robot with facial recognition or other biometric analysis at a military checkpoint to identify and report enemy combatants.	(n = 1,578)	3.907***	1.093
Q15	Use a robot with facial recognition or other biometric analysis at a U.S. military checkpoint to identify and subdue enemy combatants.	(n = 1,578)	3.586***	1.164
Q16	Use a swarm of armed autonomous drones to protect U.S. soldiers from an enemy autonomous drone swarm that is attacking.	(n = 1,570)	4.125***	0.998
Q17	Use a swarm of armed autonomous drones to preemptively destroy an enemy autonomous drone swarm.	(n = 1,575)	3.891***	1.097
Q18	Use a swarm of armed autonomous drones to attack enemy combatants.	(n = 1,577)	3.153***	1.239
Q19	Use a computer program to analyze data to identify the location of enemy targets.	(n = 1,579)	4.260***	0.919
Q20	Use a computer program to make recommendations to a military commander on how to attack targets.	(n = 1,576)	3.933***	1.036
Q21	Exploit new vulnerabilities in commercially available software to attack enemy military systems rather than notify the company of the bug.	(n = 1,577)	3.062*	1.237
Q22	Generate fake videos that show foreign leaders in compromising situations.	(n = 1,576)	1.907***	1.130
Q23	Use a U.S. fighter jet that, if it determines that the pilot is losing concentration, will seize control of the aircraft and immediately return to base.	(n = 1,581)	4.061***	1.006
Q24	Use a U.S. fighter jet that, if it determines that the pilot is losing concentration, will seize control of the aircraft and complete the mission by destroying all remaining targets.	(n = 1,579)	2.746***	1.277
Q25	Use autonomous weapons when the enemy is winning a battle without autonomous weapons.	(n = 1,578)	3.051	1.272
Q26	Use autonomous weapons when the enemy is winning a battle with autonomous weapons.	(n = 1,574)	3.818***	1.131

NOTE: We performed a one-sided t-test to show whether the mean for each question was significantly different than the midpoint of the scale. If a p-value is less than 0.05, it is flagged with one asterisk (*). If a p-value is less than 0.01, it is flagged with two asterisks (**). If a p-value is less than 0.001, it is flagged with three asterisks (***). This corresponds to confidence intervals of 95 percent, 99 percent, and 99.9 percent, respectively.

Exploratory Factor Analysis

We conducted an exploratory factor analysis (EFA) to find out if there were sets of questions (or "factors") that respondents answered similarly. A scree plot and parallel analysis both

158

suggested that there were three distinct factors worth considering. The EFA indicated that these three factors explained almost half (47 percent) of the variance.[2] The questions in the top three factors can be loosely defined as follows:

1. **General ethics of AI:** thoughts on human dignity and accountability, belief that AI will make more mistakes than humans or make war more likely, support for a ban on AWS;
2. **Scenarios that demonstrate nonlethal autonomy:** swarm force protection, data analysis for targeting, recommendations to commanders, sentry robots at checkpoints, fighter jet with a return-to-base function;
3. **Scenarios that demonstrate unconstrained autonomy:** missiles that search and destroy without human authorization, generating fake videos, fighter jet with a complete-mission function.

Demographic Analysis

Tables B.3–B.6 show how mean answers varied across the demographic groups identified in our survey results. As before, significance levels are indicated by the number of asterisks. Table B.3 shows mean survey answers based on political preference.

Table B.3. Mean Survey Answers Based on Political Preference

	Questions	Republican	Democrat	Independent	Significance
	How much do you agree with the following statements?				
Q1	War is always wrong.	2.548	3.299	3.081	***
Q2	Autonomous weapons are more likely to make mistakes than humans.	2.782	3.042	2.917	**
Q3	Removing human emotions from decisions in war is beneficial.	3.224	2.528	2.562	***
Q4	Autonomous weapons are ethically prohibited because they violate the dignity of human life.	2.967	3.465	3.297	***
Q5	Autonomous weapons are ethically prohibited because they cannot be held accountable or punished for wrongful actions.	3.255	3.611	3.636	***
Q6	The U.S. military's testing process will ensure that autonomous weapons are safe to use.	3.592	2.868	2.896	***
Q7	The United States should work with other countries to ban autonomous weapons.	3.054	3.694	3.566	***
Q8	The development of autonomous weapons will make the occurrence of wars more likely.	3.058	3.688	3.476	***
Q9	Autonomous weapons will be more accurate and precise than humans.	3.885	3.616	3.741	***

---

[2] A scree plot and parallel analysis both suggested that three factors were appropriate.

Questions	Republican	Democrat	Independent	Significance
How much do you agree that it is ethically permissible for the U.S. military to do the following?				
Q10 Continue to invest in artificial intelligence technology for military use.	4.198	3.564	3.590	***
Q11 Use missiles that autonomously search for and destroy enemy targets in war zones only if the missiles have human authorization.	4.069	3.777	3.775	***
Q12 Use missiles that autonomously search for and destroy enemy targets in war zones without human authorization.	2.622	2.078	2.145	***
Q13 Use missiles that autonomously search for and destroy enemy targets in close proximity to civilians without human authorization.	2.557	2.009	2.065	***
Q14 Use a robot with facial recognition or other biometric analysis at a military checkpoint to identify and report enemy combatants.	4.200	3.831	3.901	***
Q15 Use a robot with facial recognition or other biometric analysis at a U.S. military checkpoint to identify and subdue enemy combatants.	4.000	3.499	3.519	***
Q16 Use a swarm of armed autonomous drones to protect U.S. soldiers from an enemy autonomous drone swarm that is attacking.	4.378	4.068	4.061	***
Q17 Use a swarm of armed autonomous drones to preemptively destroy an enemy autonomous drone swarm.	4.241	3.813	3.809	***
Q18 Use a swarm of armed autonomous drones to attack enemy combatants.	3.713	2.986	3.089	***
Q19 Use a computer program to analyze data to identify the location of enemy targets.	4.489	4.231	4.181	***
Q20 Use a computer program to make recommendations to a military commander on how to attack targets.	4.164	3.896	3.863	***
Q21 Exploit new vulnerabilities in commercially available software to attack enemy military systems rather than notify the company of the bug.	3.461	2.951	3.004	***
Q22 Generate fake videos that show foreign leaders in compromising situations.	2.221	1.857	1.791	***
Q23 Use a U.S. fighter jet that, if it determines that the pilot is losing concentration, will seize control of the aircraft and immediately return to base.	4.096	4.062	4.049	***
Q24 Use a U.S. fighter jet that, if it determines that the pilot is losing concentration, will seize control of the aircraft and complete the mission by destroying all remaining targets.	3.145	2.716	2.607	***
Q25 Use autonomous weapons when the enemy is winning a battle without autonomous weapons.	3.563	2.913	2.963	***
Q26 Use autonomous weapons when the enemy is winning a battle with autonomous weapons.	4.160	3.711	3.770	***

NOTE: We performed a one-way analysis of variance (ANOVA) test to show whether the mean answers for subgroups were significantly different from each other. Lower p-values indicate higher confidence that the means are different. If a p-value is less than 0.05, it is flagged with one asterisk (*). If a p-value is less than 0.01, it is flagged with two asterisks (**). If a p-value is less than 0.001, it is flagged with three asterisks (***). This corresponds to confidence intervals of 95 percent, 99 percent, and 99.9 percent, respectively.

The results of this analysis indicate that respondents grouped by political affiliation did not answer questions in the same way. In general, Republicans were more accepting of autonomy than Democrats or Independents.

Table B.4 shows mean survey answers based on whether respondents had worked in the government or served in the military.

Table B.4. Mean Survey Answers Based on Government Work or Service in the Military

Questions	Military Experience	Government Experience	No Government or Military Experience	Significance
How much do you agree with the following statements?				
Q1 War is always wrong.	2.877	2.951	3.095	
Q2 Autonomous weapons are more likely to make mistakes than humans.	3.372	3.258	2.903	***
Q3 Removing human emotions from decisions in war is beneficial.	3.248	3.161	2.594	***
Q4 Autonomous weapons are ethically prohibited because they violate the dignity of human life.	3.298	3.268	3.323	
Q5 Autonomous weapons are ethically prohibited because they cannot be held accountable or punished for wrongful actions.	3.708	3.625	3.537	
Q6 The U.S. military's testing process will ensure that autonomous weapons are safe to use.	3.615	3.442	2.938	***
Q7 The United States should work with other countries to ban autonomous weapons.	3.450	3.553	3.530	
Q8 The development of autonomous weapons will make the occurrence of wars more likely.	3.429	3.421	3.491	
Q9 Autonomous weapons will be more accurate and precise than humans.	3.975	3.834	3.693	*
How much do you agree that it is ethically permissible for the U.S. military to do the following?				
Q10 Continue to invest in artificial intelligence technology for military use.	4.000	3.834	3.659	**
Q11 Use missiles that autonomously search for and destroy enemy targets in war zones only if the missiles have human authorization.	4.203	4.043	3.783	***
Q12 Use missiles that autonomously search for and destroy enemy targets in war zones without human authorization.	2.876	2.685	2.125	***
Q13 Use missiles that autonomously search for and destroy enemy targets in close proximity to civilians without human authorization.	2.876	2.691	2.037	***
Q14 Use a robot with facial recognition or other biometric analysis at a military checkpoint to identify and report enemy combatants.	4.286	4.069	3.875	***

161

Questions	Military Experience	Government Experience	No Government or Military Experience	Significance
Q15 Use a robot with facial recognition or other biometric analysis at a U.S. military checkpoint to identify and subdue enemy combatants.	3.983	3.847	3.541	***
Q16 Use a swarm of armed autonomous drones to protect U.S. soldiers from an enemy autonomous drone swarm that is attacking.	4.267	4.162	4.111	
Q17 Use a swarm of armed autonomous drones to preemptively destroy an enemy autonomous drone swarm.	4.252	4.050	3.862	***
Q18 Use a swarm of armed autonomous drones to attack enemy combatants.	3.567	3.478	3.100	***
Q19 Use a computer program to analyze data to identify the location of enemy targets.	4.463	4.414	4.231	**
Q20 Use a computer program to make recommendations to a military commander on how to attack targets.	4.192	4.062	3.913	**
Q21 Exploit new vulnerabilities in commercially available software to attack enemy military systems rather than notify the company of the bug.	3.911	3.562	2.981	***
Q22 Generate fake videos that show foreign leaders in compromising situations.	2.738	2.429	1.821	***
Q23 Use a U.S. fighter jet that, if it determines that the pilot is losing concentration, will seize control of the aircraft and immediately return to base.	4.149	4.092	4.057	
Q24 Use a U.S. fighter jet that, if it determines that the pilot is losing concentration, will seize control of the aircraft and complete the mission by destroying all remaining targets.	3.344	3.056	2.694	***
Q25 Use autonomous weapons when the enemy is winning a battle without autonomous weapons.	3.702	3.503	2.981	***
Q26 Use autonomous weapons when the enemy is winning a battle with autonomous weapons.	4.227	3.876	3.795	***

NOTE: We performed a one-way ANOVA test to show whether the mean answers for subgroups were significantly different from each other. Lower p-values indicate higher confidence that the means are different. If a p-value is less than 0.05, it is flagged with one asterisk (*). If a p-value is less than 0.01, it is flagged with two asterisks (**). If a p-value is less than 0.001, it is flagged with three asterisks (***). This corresponds to confidence intervals of 95 percent, 99 percent, and 99.9 percent, respectively.

As the table indicates, respondents who previously served in the military or worked in government believed more than those who did not that autonomous weapons were more likely to make mistakes than humans. However, former military members were also significantly more confident in the military's testing process and supported continued investment in military AI.

Table B.5 shows mean survey answers based on age.

Table B.5. Mean Survey Answers Based on Age

	Questions	<30 Years	30–49 Years	>50 Years	Significance
	How much do you agree with the following statements?				
Q1	War is always wrong.	3.245	2.994	2.933	***
Q2	Autonomous weapons are more likely to make mistakes than humans.	2.87	2.974	3.018	
Q3	Removing human emotions from decisions in war is beneficial.	2.713	2.616	2.759	
Q4	Autonomous weapons are ethically prohibited because they violate the dignity of human life.	3.395	3.324	3.062	**
Q5	Autonomous weapons are ethically prohibited because they cannot be held accountable or punished for wrongful actions.	3.609	3.530	3.445	
Q6	The U.S. military's testing process will ensure that autonomous weapons are safe to use.	2.990	2.996	3.088	
Q7	The United States should work with other countries to ban autonomous weapons.	3.556	3.545	3.379	
Q8	The development of autonomous weapons will make the occurrence of wars more likely.	3.698	3.410	3.295	***
Q9	Autonomous weapons will be more accurate and precise than humans.	3.820	3.664	3.671	*
	How much do you agree that it is ethically permissible for the U.S. military to do the following?				
Q10	Continue to invest in artificial intelligence technology for military use.	3.538	3.710	3.973	***
Q11	Use missiles that autonomously search for and destroy enemy targets in war zones only if the missiles have human authorization.	3.827	3.788	3.934	
Q12	Use missiles that autonomously search for and destroy enemy targets in war zones without human authorization.	2.192	2.173	2.313	
Q13	Use missiles that autonomously search for and destroy enemy targets in close proximity to civilians without human authorization.	2.122	2.086	2.273	
Q14	Use a robot with facial recognition or other biometric analysis at a military checkpoint to identify and report enemy combatants.	3.776	3.944	4.110	***
Q15	Use a robot with facial recognition or other biometric analysis at a U.S. military checkpoint to identify and subdue enemy combatants.	3.434	3.615	3.85	***
Q16	Use a swarm of armed autonomous drones to protect U.S. soldiers from an enemy autonomous drone swarm that is attacking.	4.075	4.105	4.309	**
Q17	Use a swarm of armed autonomous drones to preemptively destroy an enemy autonomous drone swarm.	3.786	3.904	4.093	**

163

Questions		<30 Years	30–49 Years	>50 Years	Significance
Q18	Use a swarm of armed autonomous drones to attack enemy combatants.	3.078	3.142	3.367	*
Q19	Use a computer program to analyze data to identify the location of enemy targets.	4.153	4.283	4.427	***
Q20	Use a computer program to make recommendations to a military commander on how to attack targets.	3.844	3.944	4.115	**
Q21	Exploit new vulnerabilities in commercially available software to attack enemy military systems rather than notify the company of the bug.	2.996	3.058	3.233	
Q22	Generate fake videos that show foreign leaders in compromising situations.	1.927	1.835	2.119	**
Q23	Use a U.S. fighter jet that, if it determines that the pilot is losing concentration, will seize control of the aircraft and immediately return to base.	4.047	4.074	4.053	
Q24	Use a U.S. fighter jet that, if it determines that the pilot is losing concentration, will seize control of the aircraft and complete the mission by destroying all remaining targets.	2.737	2.716	2.863	
Q25	Use autonomous weapons when the enemy is winning a battle without autonomous weapons.	2.958	3.04	3.286	**
Q26	Use autonomous weapons when the enemy is winning a battle with autonomous weapons.	3.775	3.818	3.916	

NOTE: We performed a one-way ANOVA test to show whether the mean answers for subgroups were significantly different from each other. Lower p-values indicate higher confidence that the means are different. If a p-value is less than 0.05, it is flagged with one asterisk (*). If a p-value is less than 0.01, it is flagged with two asterisks (**). If a p-value is less than 0.001, it is flagged with three asterisks (***). This corresponds to confidence intervals of 95 percent, 99 percent, and 99.9 percent, respectively.

The analysis indicates that, in general, younger respondents were significantly warier of using autonomy than older respondents.

Table B.6 shows mean survey answers based on gender.

Table B.6. Mean Survey Answers Based on Sex

Questions		Male	Female	Significance
	How much do you agree with the following statements?			
Q1	War is always wrong.	3.002	3.149	*
Q2	Autonomous weapons are more likely to make mistakes than humans.	2.896	3.025	*
Q3	Removing human emotions from decisions in war is beneficial.	2.814	2.49	***
Q4	Autonomous weapons are ethically prohibited because they violate the dignity of human life.	3.173	3.472	***
Q5	Autonomous weapons are ethically prohibited because they cannot be held accountable or punished for wrongful actions.	3.492	3.618	*
Q6	The U.S. military's testing process will ensure that autonomous weapons are safe to use.	3.065	2.944	*

	Questions	Male	Female	Significance
Q7	The United States should work with other countries to ban autonomous weapons.	3.419	3.657	***
Q8	The development of autonomous weapons will make the occurrence of wars more likely.	3.433	3.553	*
Q9	Autonomous weapons will be more accurate and precise than humans.	3.890	3.474	***
	How much do you agree that it is ethically permissible for the U.S. military to do the following?			
Q10	Continue to invest in artificial intelligence technology for military use.	3.822	3.525	***
Q11	Use missiles that autonomously search for and destroy enemy targets in war zones only if the missiles have human authorization.	3.914	3.708	***
Q12	Use missiles that autonomously search for and destroy enemy targets in war zones without human authorization.	2.367	1.990	***
Q13	Use missiles that autonomously search for and destroy enemy targets in close proximity to civilians without human authorization.	2.259	1.957	***
Q14	Use a robot with facial recognition or other biometric analysis at a military checkpoint to identify and report enemy combatants.	4.054	3.725	***
Q15	Use a robot with facial recognition or other biometric analysis at a U.S. military checkpoint to identify and subdue enemy combatants.	3.699	3.446	***
Q16	Use a swarm of armed autonomous drones to protect U.S. soldiers from an enemy autonomous drone swarm that is attacking.	4.23	3.994	***
Q17	Use a swarm of armed autonomous drones to preemptively destroy an enemy autonomous drone swarm.	4.012	3.746	***
Q18	Use a swarm of armed autonomous drones to attack enemy combatants.	3.325	2.938	***
Q19	Use a computer program to analyze data to identify the location of enemy targets.	4.357	4.141	***
Q20	Use a computer program to make recommendations to a military commander on how to attack targets.	4.062	3.777	***
Q21	Exploit new vulnerabilities in commercially available software to attack enemy military systems rather than notify the company of the bug.	3.322	2.733	***
Q22	Generate fake videos that show foreign leaders in compromising situations.	2.091	1.672	***
Q23	Use a U.S. fighter jet that, if it determines that the pilot is losing concentration, will seize control of the aircraft and immediately return to base.	4.12	3.988	*
Q24	Use a U.S. fighter jet that, if it determines that the pilot is losing concentration, will seize control of the aircraft and complete the mission by destroying all remaining targets.	2.85	2.623	***
Q25	Use autonomous weapons when the enemy is winning a battle without autonomous weapons.	3.27	2.778	***
Q26	Use autonomous weapons when the enemy is winning a battle with autonomous weapons.	3.946	3.656	***

NOTE: We performed a one-way ANOVA test to show whether the mean answers for subgroups were significantly different from each other. Lower p-values indicate higher confidence that the means are different. If a p-value is less than 0.05, it is flagged with one asterisk (*). If a p-value is less than 0.01, it is flagged with two asterisks (**). If a p-value is less than 0.001, it is flagged with three asterisks (***). This corresponds to confidence intervals of 95 percent, 99 percent, and 99.9 percent, respectively.

This analysis suggests that men are much more accepting of autonomy than women.

Open-Ended Responses

At the end of the survey, respondents had an option to provide additional thoughts or comments on the ethical implications of integrating AI into the U.S. military and AI's subsequent impact on the character of war. Many took this opportunity quite seriously. Over 30 percent of respondents wrote more than 100 characters, indicating a sign of high engagement and interest in the survey.

We used an in-house text analytics tool called RAND-Lex to parse and collate these free-form answers.[3] Text analytics is a process used to glean high-quality information from text. The pattern recognition techniques applied in text analytics typically generate results that possess some degree of novelty, relevance, or interest to the user.[4]

We performed a collocation analysis to find two- and three-word phrases within seven-word windows that were uncommonly prevalent. To do this, RAND-Lex uses an association metric that scores these phrases based on their frequency and rarity. Table B.7 shows a list of the top 20 phrases that respondents mentioned in the open-ended response section.

Table B.7. Collocation Analysis of Open-Ended Survey Responses

Frequency	Score	Token 1	Token 2	Token 3
5	16.2342	no	matter	how
8	13.96368	without	human	intervention
8	13.11568	without	human	authorization
11	12.87384	a	good	thing
7	12.7097	a	bad	idea
13	11.98459	I	don't	know
8	11.70071	a	good	idea
8	9.20783	slippery	slope	
6	8.50328	held	accountable	
6	7.79279	fine	line	
5	7.07032	save	American	
6	6.8859	certain	situations	
5	6.84793	full	control	
7	6.35222	civilian	casualties	
17	6.12314	save	lives	

---

[3] For more on RAND-Lex, see Doug Irving, "Big Data, Big Questions," *RAND Review*, October 16, 2017.

[4] William H. Dutton and Paul W. Jeffreys, *World Wide Research: Reshaping the Sciences and Humanities*, Cambridge, Mass.: MIT Press, 2010.

Frequency	Score	Token 1	Token 2	Token 3
8	6.11097	life	death	
5	5.67177	mistakes	made	
7	5.62287	death	decisions	
8	5.31436	more	accurate	
7	5.14345	decision	making	

The collocation analysis in Table B.7 indicates a tone of caution and concern in the open-ended answers. Phrases such as "without human intervention/authorization," "full control," and "death decisions" suggest that the human-in-the-loop idea is especially important to respondents. "Save lives," "save American," and "more accurate" seem more positive in tone, whereas "slippery slope" and "fine line" reflect a wariness about having proper measures in place to restrain military AI.

Next, we took a deeper look at four of the words found in the collocation analysis by examining how they were used in context. Table B.8 shows five instances of each of the four target words ("accountable," "civilian," "slope," "decision") as they were used in open-ended responses.

Table B.8. How Certain Target Words in Open-Ended Responses Were Used in Context

Left	Target	Right
We need to be held	accountable	for our actions, not create easier ways to bomb others.
This way no one will ever be held	accountable	Is that what we want?
. . .would rather have people held responsible and	accountable	for any and all action. This will be extremely difficult.
. . .because they don't think for themselves and cannot be held	accountable	for their actions. No. It's scary and tough to think…
I think the biggest thing is someone needs to be	accountable	for failure.
If the software relating to acceptable	civilian	risk is consistent with policy, then the autonomous weapons can . . .
Ensure that it is done in war zones rather than near	civilian	populations. I think that there are some ethical considerations that
Usually human lives are killed, both military and	civilian	during war. At some point people need to consider this.
. . .precise, thus reducing the cost and lowering the risk for	civilian	casualties. It also limits the exposure of U.S. soldiers to . . .
. . . to have very strict rules of engagement which would reduce	civilian	casualties. I think there are certain dangers around AI and . . .
I think it's a slippery	slope	when you include AI that can operate without human interaction . . .
Need to be wary of a slippery	slope	toward potential abuses of technology on civilians.

Left	Target	Right
Just have to say that it is quite the slippery	slope	and could end up creating an uncontrollable ever-escalating situation that . . .
I guess it is a slippery	slope	We as humans are not very good at predicting all
Using AI seems like a slippery	slope	especially in terms of not having human authorization.
AI should be used as a tool to enhance human	decision	making and help suggest clearer and better strategies but ultimate . . .
. . . always be the ability for a human to override any	decision	made by AI because sometimes there are rare and extenuating . . .
Removing humans from the process and	decision	making makes it too easy to remove humans from blame.
In war scenarios humans should always have ultimate control and	decision	making ability. I think we need humans making these decisions . . .
. . . ways to integrate AI in the military but the final	decision	on missions and targeting must remain with humans.

These four target words identify four major concerns that the respondents had with military AI. The first concern is that autonomous systems will not be as safe because they cannot be held accountable for their actions. Respondents would prefer for someone to be held responsible when mistakes are made.

The next concern is related to the risk military AI poses to civilian populations. Respondents are generally opposed to putting the civilian population at risk, but there is an acknowledgment that certain levels of collateral damage are acceptable to increase the safety of soldiers and the effectiveness of the mission.

"Slippery slope" was used multiple times to express concern that using AI in the military will eventually lead to undesired escalation dynamics, abuse, and unpredicted behaviors.

Finally, the word "decision" reiterates the ubiquitous concern about granting full decisionmaking authority to a machine. Once again, this illustrates how critical the "human in the loop" idea is to people.

References

"2018 Group of Governmental Experts on Lethal Autonomous Weapons Systems (LAWS)," United Nations Office at Geneva, 2018. As of August 30, 2018: https://www.unog.ch/__80256ee600585943.nsf/(httpPages)/7c335e71dfcb29d1c1258243003 e8724?OpenDocument&ExpandSection=3#_Section3

"303060302-空军装备预研创新-基于自动图像识别的发动机叶片裂纹检测研究" ["303060302-Air Force Equipment Research and Innovation: Engine Blade Crack Detection with Autonomous Image Recognition"], 全军武器装备采购信息网 [*Whole Military Weapons and Equipment Purchase Information Net*], August 10, 2017.

Air Force Research Laboratory /RQQA, "Air Force Research Laboratory Test and Evaluation, Verification and Validation of Autonomous Systems Challenge Exploration Final Report," *Final Report*, November 13, 2014. As of June 29, 2018: http://www.dtic.mil/get-tr-doc/pdf?AD=ADA614199

Alderman, Daniel, and Johnathan Ray, "Best Frenemies Forever: Artificial Intelligence, Emerging Technologies, and China-US Strategic Competition," SITC Research Briefs, Series 9, January 10, 2017. As of July 3, 2018: https://escholarship.org/uc/item/2pq268gz

Allen, Gregory C., "Project Maven Brings AI to the Fight Against ISIS," *Bulletin of the Atomic Scientists*, December 21, 2017. As of June 28, 2018: https://thebulletin.org/project-maven-brings-ai-fight-against-isis11374

Amodei, Dario, Chris Olay, Jacob Steinhardt, Paul Christiano, John Schulman, and Dan Mané, "Concrete Problems in AI Safety," July 25, 2016. As of September 10, 2018: https://arxiv.org/abs/1606.06565

"An Open Source Machine Learning Framework for Everyone," TensorFlow, n.d. As of September 10, 2018: https://www.tensorflow.org

Arms Control Association, *Facts Sheets and Briefs: Chemical Weapons Conventions and States-Parties*, June 2018. As of July 2, 2018: https://www.armscontrol.org/factsheets/cwcsig

" 陆军预研-0243-无人机多机自主协同技术" ["Army Research-0243: Technology for the Autonomous Cooperation for Multiple Drones"], 全军武器装备采购信息网 [*Whole Military Weapons and Equipment Purchase Information Net*], July 27, 2016.

Article 36, "Autonomous Weapon Systems: Evaluating the Capacity for Meaningful Human Control in Weapon Review Processes," Discussion paper for the Convention on Certain Conventional Weapons (CCW) Group of Governmental Experts meeting on Lethal Autonomous Weapons Systems (LAWS), November 2017. As of July 23, 2019:
http://www.article36.org/wp-content/uploads/2013/06/Evaluating-human-control-1.pdf

"Article 36 Reviews and Addressing Lethal Autonomous Weapons Systems," Article36, April 11–15, 2016, p. 2. As of July 06, 2018:
http://www.article36.org/wp-content/uploads/2016/04/LAWS-and-A36.pdf

"Iskusstvennyy Intellekt Nauchilsya Privlekat' Gospodderzhku" ["Artificial Intelligence Has Learned to Attract State Support"], *Коммерсант* [*Kommersant*], April 4, 2017. As of July 6, 2018:
https://www.kommersant.ru/doc/3260988

"Autonomy in Weapon Systems," U.S. submission to the Meeting of the Group of Governmental Experts of the High Contracting Parties to the Convention on Prohibitions or Restrictions on the Use of Certain Conventional Weapons Which May Be Deemed to Be Excessively Injurious or to Have Indiscriminate Effects, Geneva, November 10, 2017. As of August 22, 2019:
https://www.unog.ch/80256EDD006B8954/(httpAssets)/99487114803FA99EC12581D4006 5E90A/$file/2017_GGEonLAWS_WP6_USA.pdf

Avdeev, Yuri, "На Пути К Искусственному Интеллекту" ["Toward Artificial Intelligence"], *Красная Звезда* [*Red Star*] March 16, 2018. As of July 6, 2018:
http://archive.redstar.ru/index.php/2011-07-25-15-57-07/item/36511-na-puti-k-iskusstvennomu -intellektu?attempt=1

Axe, David, "That Time an Air Force F-16 and an Army Missile Battery Fought Each Other," *Medium*, July 5, 2015. As of June 27, 2018:
https://medium.com/war-is-boring/that-time-an-air-force-f-16-and-an-army-missile-battery -fought-each-other-bb89d7d03b7d

BAE Systems, "Taranis," n.d. As of September 1, 2018:
https://www.baesystems.com/en/product/taranis

Baraniuk, Chris, "AI Fighter Pilot Wins in Combat Simulation," *BBC News*, June 28, 2016: As of February 4, 2019:
https://www.bbc.com/news/technology-36650848

Barbuk, Sergei, "ФПИ Предложил Минобороны Стандарты Для Искусственного Интеллект" ["FPI Offered the Ministry of Defense Standards for Artificial Intelligence"], *Новости ВПК* [*Novosti VPK*], March 20, 2018. As of July 06, 2018:
https://www.vpk-news.ru/news/41794

Bartlett, Matt, "China's Game of Drones," Australia Institute of International Affairs, April 11, 2018. As of July 2, 2018:
http://www.internationalaffairs.org.au/australianoutlook/china-game-drones/

"BBQ-905 型激光压制干扰器" ["BBQ-905 Laser Suppressor and Disturber"], *百度百科* [*Baidu Encyclopedia*], January 23, 2018. As of July 2, 2018:
https://baike.baidu.com/item/BBQ-905%E5%9E%8B%E6%BF%80%E5%85%89%E5%8E%8B%E5%88%B6%E5%B9%B2%E6%89%B0%E5%99%A8/18887158

Bendett, Samuel, "Red Robots Rising: Behind the Rapid Development of Russian Unmanned Military Systems," *Strategy Bridge*, December 12, 2017. As of July 6, 2018:
https://thestrategybridge.org/the-bridge/2017/12/12/red-robots-rising-behind-the-rapid-development-of-russian-unmanned-military-systems

———, "The Russian Military Wants Students to Design Its New Underwater Drone," *War Is Boring*, March 7, 2018. As of July 6, 2018:
https://warisboring.com/the-russian-military-wants-students-to-design-its-new-underwater-drone/

Bennetts, Marc, "Soviet Officer Who Averted Cold War Nuclear Disaster Dies Aged 77," *The Guardian*, September 18, 2017. As of February 26, 2019:
https://www.theguardian.com/world/2017/sep/18/soviet-officer-who-averted-cold-war-nuclear-disaster-dies-aged-77

"网民科普阅读大数据报告显示 人工智能成最热话题" ["Big Data Report on the Reading Habits in Popular Science Articles Show AI is the Hottest Topic"], *Xinhua*, March 28, 2018. As of July 3, 2018:
http://www.xinhuanet.com/book/2018-03/30/c_129841263.htm

"基于大数据的生产过程仿真和控制系统" ["A Big Data System to Simulate and Control Production"], *全军武器装备采购信息网* [*Whole Military Weapons and Equipment Purchase Information Net*], November 22, 2016. As of July 2, 2018:
http://www.weain.mil.cn/cgcms/jppt/jppt/queryDetailInfo?SiteID=122&ID=15154&CatalogInner=001302&id=538925&ssddwid=7

Blinde, Loren, "US Air Force, Lockheed Martin, Demonstrate Manned/Unmanned Teaming," *Intelligence Community News*, April 11, 2017. As of June 28, 2018:
http://intelligencecommunitynews.com/us-air-force-lockheed-martin-demonstrate-mannedunmanned-teaming/

Bostrom, Nick, *Superintelligence: Paths, Dangers, Strategies*, Oxford, UK: Oxford University Press, 2014.

Bostrom, Nick, and Milan M. Ćirković, eds., *Global Catastrophic Risks*, New York: Oxford University Press, 2011.

Boulanin, Vincent, "The Promise and Perils of Artificial Intelligence for Nuclear Stability," *Our World*, December 7, 2018. As of January 30, 2019:
https://ourworld.unu.edu/en/the-promise-and-perils-of-artificial-intelligence-for-nuclear-stability

Boulanin, Vincent, and Maaike Verbruggen, "Mapping the Development of Autonomy in Unmanned Systems," Stockholm International Peace Research Institute, November 2017, p. 43. As of July 6, 2018:
https://www.sipri.org/sites/default/files/2017-11/siprireport_mapping_the_development_of_autonomy_in_weapon_systems_1117_1.pdf

Bowden, Lord, "The Story of IFF (Identification Friend or Foe)," *IEE Proceedings A (Physical Science, Measurement and Instrumentation, Management and Education, Reviews),* Vol. 132, No. 6, 1985, pp. 435–437.

Brands, Hal, "Paradoxes of the Gray Zone," Foreign Policy Research Institute, February 5, 2016. As of September 2, 2018:
https://www.fpri.org/article/2016/02/paradoxes-gray-zone/

Browne, Ryan, "US General Warns of Out-of-Control Killer Robots," *CNN Politics*, July 18, 2017. As of June 29, 2018:
https://www.cnn.com/2017/07/18/politics/paul-selva-gary-peters-autonomous-weapons-killer-robots/index.html

Busby, Mattha, and Anthony Cuthbertson, "Killer Robots Ban Blocked by U.S. and Russia at UN Meeting," *The Independent*, September 3, 2018. As of January 31, 2019:
https://www.independent.co.uk/life-style/gadgets-and-tech/news/killer-robots-un-meeting-autonomous-weapons-systems-campaigners-dismayed-a8519511.html

Cambria, Erik, "Affective Computing and Sentiment Analysis," *IEEE Intelligent Systems*, Vol. 31, No. 2, March 2016, pp. 102–107.

"Campaign to Stop Killer Robots: The Solution," *ST THOMAS AQUINAS VERSUS NASA*, March 10, 2018. As of August 22, 2019:
https://www.stthomasaquinasversusnasa.com/2018/03/campaign-to-stop-killer-robots-solution.html

Carpenter, Charli, "How Do Americans Feel About Fully Autonomous Weapons?" Duck of Minerva, June 19, 2013. As of July 17, 2019:
http://duckofminerva.com/2013/06/how-do-americans-feel-about-fully-autonomous-weapons.html

曾行贱 [Ceng Xingjian] and 邵婧 [Shao Jing], "海军柳州舰组织实战背景高难科目训练" ["Warship Liuzhou Undergoes Realistic Training"], 中国海军网 [*China Navy Net*], January 11, 2018. As of July 3, 2018:
http://www.mod.gov.cn/power/2018-01/11/content_4802229.htm

Chen, Frank, "China Shows Off Drone Brigade at Guangzhou Fortune Forum Gala," *Asia Times*, December 8, 2017.

Chen, Stephen, "'Laser AK-47'? Chinese Developer Answers Sceptics with Videos Of Gun Being Tested," *South China Morning Post*, July 5, 2018. As of July 11, 2018:
https://www.scmp.com/news/china/diplomacy-defence/article/2153791/laser-ak-47-chinese-developer-answers-sceptics-videos

"China—Air Force," *Jane's World Air Forces*, March 15, 2018.

"China: Ratify Key International Human Rights Treaty," Human Rights Watch, October 8, 2013. As of July 2, 2018:
https://www.hrw.org/news/2013/10/08/china-ratify-key-international-human-rights-treaty

中国中央电视台[China Central Television], "海斗号：我国首台万米水下机器人" ["Sea Challenger: Our Nation's First Underwater Robot to Dive 1,000 Meters"], 爱奇艺 [*Iqiyi*], August 23, 2016. As of July 2, 2018:
http://www.iqiyi.com/v_19rrly3kcc.html

中国电子技术标准化研究院 [China Electronic Standardization Institute], 人工智能标准化白皮书 [*AI Standardization White Paper*], 2018. As of July 3, 2018:
http://www.sgic.gov.cn/upload/f1ca3511-05f2-43a0-8235-eeb0934db8c7/20180122/5371516606048992.pdf

"China Has Turned Xinjiang into a Police State like No Other," *The Economist*, May 31, 2018.

"China Unveils GL5 Active Protection System for Main Battle Tanks," *Defense Blog*, August 16, 2017. As of July 2, 2018:
http://defence-blog.com/army/china-unveils-gl5-active-protection-system-for-main-battle-tanks.html

"China's CSSC Unveiled the Type 730C Dual Gun and Missile CIWS," *Navy Recognition*, March 9, 2017. As of July 2, 2018:
http://www.navyrecognition.com/index.php/news/naval-exhibitions/2017/navdex-2017-show-daily-news/4970-china-s-cssc-unveiled-the-type-730c-dual-gun-and-missile-ciws.html

Chivvis, Christopher S., "Understanding Russian 'Hybrid Warfare' and What Can Be Done About It: Addendum," Santa Monica, Calif.: RAND Corporation, CT-468/1, 2017. As of July 6, 2018:
https://www.rand.org/pubs/testimonies/CT468z1.html

Clausewitz, Carl, *On War*, ed. and trans. Michael Howard and Peter Paret. Princeton, N.J.: Princeton University Press, 1976.

Coldeway, Devin, "Carnegie Mellon's Mayhem AI Takes Home $2 Million from DARPA's Cyber Grand Challenge," *TechCrunch*, August 5, 2016. As of June 28, 2018:
https://techcrunch.com/2016/08/05/carnegie-mellons-mayhem-ai-takes-home-2-million-from-darpas-cyber-grand-challenge/

"Combat Robots for Russian Troops to Go into Serial Production This Year—Defense Minister," *TACC* [*TASS*], March 15, 2018. As of July 6, 2018:
http://tass.com/defense/994310

"Combat Tests in Syria Brought to Light Deficiencies of Russian Unmanned Mini-Tank," *Defense Blog*, June 18, 2018. As of July 6, 2018:
https://defence-blog.com/army/combat-tests-syria-brought-light-deficiencies-russian -unmanned-mini-tank.html

Conger, Kate, "The Fight for a Massive Pentagon Cloud Contract Is Heating Up," *Gizmodo*, May 8, 2018. As of June 28, 2018:
https://gizmodo.com/the-fight-for-a-massive-pentagon-cloud-contract-is-heat-1825517332

"The Cooperative Engagement Capability," *Johns Hopkins APL Technical Digest*, Vol. 16, No. 4 (1995), pp. 377–396. As of February 8, 2019:
https://www.jhuapl.edu/techdigest/TD/td1604/APLteam.pdf

Cronbach, L. J., "Coefficient Alpha and the Internal Structure of Tests," *Psychometrika*, Vol. 16, 1951, pp. 297–334.

Cummings, Mary L., "Creating Moral Buffers in Weapon Control Interface Design," *IEEE Technology and Society Magazine*, Vol. 23, No. 3, Fall 2004, pp. 28–33. As of September 5, 2018:
https://ieeexplore.ieee.org/document/1337888/

———, "Automation Bias in Intelligent Time Critical Decision Support Systems," *AIAA 1st Intelligent Systems Technical Conference*, 2004, published online June 19, 2012. As of September 5, 2018:
https://arc.aiaa.org/doi/abs/10.2514/6.2004-6313

Danzig, Richard, "Technology Roulette: Managing Loss of Control as Many Militaries Pursue Technological Superiority," Center for a New American Security, June 2018. As of September 10, 2018:
https://s3.amazonaws.com/files.cnas.org/documents/CNASReport-Technology-Roulette -DoSproof2v2.pdf?mtime=20180628072101

"业务导向的大数据挖掘分析" ["Data Mining and Analysis for Business"], 全军武器装备采购信息网 [*Whole Military Weapons and Equipment Purchase Information Net*], May 12, 2017.

Defense Advanced Research Projects Agency, *The DARPA Grand Challenge: Ten Years Later*, Washington, D.C., March 13, 2014. As of August 29, 2018:
https://www.darpa.mil/news-events/2014-03-13

Dent, Steve, "Autonomous Helicopter Makes First Operational Delivery to Marines," *Engadget*, May 18, 2018. As of June 28, 2018:
https://www.engadget.com/2018/05/18/aurora-aacus-autonomous-helicopter-supply-delivery/

DeSimone, Antonio, and Nicholas Horton, *Sony's Nightmare Before Christmas: The 2014 North Korean Cyber Attack on Sony and Lessons for US Government Actions in Cyberspace*, National Security Report, Johns Hopkins Applied Physics Laboratory, Laurel, Md., 2017. As of January 29, 2019:
https://www.jhuapl.edu/Content/documents/SonyNightmareBeforeChristmas.pdf

DeSutter, Paula A., "China's Record of Proliferation Activities," Testimony before the United States–China Commission, Washington, D.C., July 24, 2003. As of July 2, 2018:
https://2001-2009.state.gov/t/vci/rls/rm/24518.htm

DeVellis, R. F., *Scale Development: Theory and Applications*, 2nd ed., Thousand Oaks, Calif.: Sage Publications, 2003.

狄伯文 [Di Bowen], 赵建文 [Zhao Jianwen], and 钱晓虎 [Qian Xiaohu], "打赢明天战争：信息化武器装备+智能化创新步伐" ["Winning Tomorrow's Wars: Informationized Weapons and Equipment + Intelligentized Innovation Marches On"], 中国军网综合 [*China Military Online*], November 24, 2017. As of July 12, 2018:
http://www.81.cn/2017zt/2017-11/24/content_7840570_3.htm

Dierking, Phil, "Support Grows for a Treaty to Ban Killer Robots," *VOA Learning English*, August 30, 2018. As of January 31, 2019:
https://learningenglish.voanews.com/a/support-grows-for-a-treaty-to-ban-killer-robots/4551083.html

Ding, Jeffrey, Paul Triolo, and Samm Sacks, "Chinese Interests Take a Big Seat at the AI Governance Table," Center for New American Security, June 20, 2018.

Dobkin, Adin, "DoD Maven AI Project Develops First Algorithms, Starts Testing," Defense Systems, November 3, 2017. As of June 28, 2018:
https://defensesystems.com/articles/2017/11/03/maven-dod.aspx

DoD—*See* U.S. Department of Defense.

Dorosin, Josh, "U.S. Delegation Statement on Human-Machine Interaction," The Meeting of the Group of Governmental Experts of the High Contracting Parties to the CCW on Lethal Autonomous Weapons Systems, August 2018b. As of September 5, 2018:
https://geneva.usmission.gov/2018/08/28/u-s-delegation-statement-on-human-machine-interaction/

———, "U.S. Delegation Statement on Possible Options," The Meeting of the Group of Governmental Experts of the High Contracting Parties to the CCW on Lethal Autonomous Weapons Systems, August 2018a. As of September 5, 2018:
https://geneva.usmission.gov/2018/08/29/u-s-delegation-statement-on-possible-policy-options/

Dougherty, Jill, and Molly Jay, "Russia Tries to Get Smart About Artificial Intelligence," *Wilson Quarterly*, Spring 2018. As of July 6, 2018:
https://www.wilsonquarterly.com/quarterly/living-with-artificial-intelligence/russia-tries-to-get-smart-about-artificial-intelligence/

杜严勇 [Du Yanyong], "现代军用机器人的伦理困境" ["The Ethical Difficulty of Modern Military Robots"], 伦理学研究 [*Studies in Ethics*], No. 5, 2014. As of July 3, 2018:
http://kns55.en.eastview.com/kcms/detail/detail.aspx?recid=&FileName=YJLL201405020&DbName=CJFD2014&DbCode=CJFD&uid=WEE2cU9OY1hxNUMvQkljPQ,
As of July 3, 2018, also available at
http://www.doc88.com/p-8661279514294.html

Dunlap, Charles, "Accountability and Autonomous Weapons: Much Ado About Nothing?" *Temple International and Comparative Law Journal*, Vol. 30, No. 1, 2016, pp. 63–76. As of September 5, 2018:
https://sites.temple.edu/ticlj/files/2017/02/30.1.Dunlap-TICLJ.pdf

Dutton, William H., and Paul W. Jeffreys, *World Wide Research: Reshaping the Sciences and Humanities*, Cambridge, Mass.: MIT Press, 2010.

Eberstadt, Nicholas, "Demography and Human Resources: Unforgiving Constraints for a Russia in Decline," Jamestown Foundation, September 13, 2016. As of July 6, 2018:
https://jamestown.org/program/demography-human-resources-unforgiving-constraints-russia-decline/

Edwards, Scott, and Steven Livingston, "Fake News Is About to Get a Lot Worse. That Will Make It Easier to Violate Human Rights—and Get Away with It," *Washington Post*, April 3, 2018. As of July 6, 2018:
https://www.washingtonpost.com/news/monkey-cage/wp/2018/04/03/fake-news-is-about-to-get-a-lot-worse-that-will-make-it-easier-to-violate-human-rights-and-get-away-with-it/?noredirect=on&utm_term=.be97cd7a6f59

"Elements Supporting the Prohibition of Lethal Autonomous Weapons Systems," working paper submitted by the Holy See to the Group of Governmental Experts of the High Contracting Parties to the Convention on Prohibitions or Restrictions on the Use of Certain Conventional Weapons Which May Be Deemed to Be Excessively Injurious or to Have Indiscriminate Effects, April 7, 2016. As of September 11, 2018:
https://www.unog.ch/80256EDD006B8954/(httpAssets)/752E16C02C9AECE4C1257F8F0040D05A/$file/2016_LAWSMX_CountryPaper_Holy+See.pdf

"Emerging Commonalities, Conclusions and Recommendations," Group of Governmental Experts of the High Contracting Parties to the Convention on Prohibitions or Restrictions on the Use of Certain Conventional Weapons Which May Be Deemed to Be Excessively Injurious or to Have Indiscriminate Effects, Geneva, August 2018. As of August 22, 2019:

176

https://www.unog.ch/80256EDD006B8954/(httpAssets)/EB4EC9367D3B63B1C12582FD00 57A9A4/$file/GGE+LAWS+August_EC,+C+and+Rs_final.pdf

Engstrom, Jeffrey, *Systems Confrontation and System Destruction Warfare: How the Chinese People's Liberation Army Seeks to Wage Modern Warfare*, Santa Monica, Calif.: RAND Corporation, RR1708, 2018. As of July 21, 2018: https://www.rand.org/pubs/research_reports/RR1708.html

"中國這一武器雖然起步晚， 但連美國都忌憚" ["Even Though China Is a Latecomer to These Weapons, It Still Gives America Pause"], 每日头条 [*Meiri Toutiao*], August 9, 2017. As of July 2, 2018: https://kknews.cc/military/xr4ppk9.html

Evtimov, Ivan, Kevin Eykholt, Earlence Fernandes, and Bo Li, "Physical Adversarial Examples Against Deep Neural Networks," *Berkeley Artificial Intelligence Research*, December 2017. As of September 5, 2018: http://bair.berkeley.edu/blog/2017/12/30/yolo-attack/

Executive Services Directorate (ESD), DoD Directive 3000.09, *Autonomy in Weapon Systems*, November 21, 2012, revised May 8, 2017. As of August 30, 2018: http://www.esd.whs.mil/Portals/54/Documents/DD/issuances/dodd/300009p.pdf

Falk, Emily, and Michael Platt, "What Your Facebook Network Reveals About How You Use Your Brain," *Scientific American*, July 9, 2018. As of August 5, 2018: https://blogs.scientificamerican.com/observations/what-your-facebook-network-reveals -about-how-you-use-your-brain/

Federation of American Scientists, Defense Science Board, *Summer Study on Autonomy*, June 2016. As of September 1, 2018: https://fas.org/irp/agency/dod/dsb/autonomy-ss.pdf

Fein, Geoff, "DARPA's XAI Seeks Explanations from Autonomous Systems," *Janes*, November 16, 2017.

Fischer, Benjamin B., "A Cold War Conundrum: The 1983 Soviet War Scare," *Library*, Central Intelligence Agency, July 7, 2008. As of August 23, 2019: https://www.cia.gov/library/center-for-the-study-of-intelligence/csi-publications/books-and -monographs/a-cold-war-conundrum/source.htm

Forum on Leadership, "Closing Conversation with Jeff Bezos, Co-Presented with SMU," Dallas, Tex.: George W. Bush Presidential Center, 2018. As of August 30, 2018: https://www.bushcenter.org/takeover/sessions/forum-leadership/bezos-closing-conversation .html

Foss, Christopher F., "Norinco Details New Land-Based CIWS," *Jane's International Defense Review*, March 3, 2015.

Freedberg, Sydney J. Jr., "Tomahawk vs LRASM: Raytheon Gets $119M for Anti-Ship Missile," *Breaking Defense*, September 11, 2017. As of June 26, 2018: https://breakingdefense.com/2017/09/tomahawk-vs-lrasm-raytheon-gets-119m-for-anti-ship -missile/

Frisk, Adam, "What Is Project Maven? The Pentagon AI Project Google Employees Want Out Of," *Global News*, April 5, 2018. As of August 30, 2018: https://globalnews.ca/news/4125382/google-pentagon-ai-project-maven/

"基金-61401370502-基于深度学习的信号特征提取技术研究" ["Fund-60401370502: Technology That Uses Deep Learning to Identify Signal Characteristics"], 全军武器装备采购信息网 [*Whole Military Weapons and Equipment Purchase Information Net*], August 1, 2016.

"海军创新-30201050110-基于深度学习的海洋遥感目标信息挖掘技术" ["Fund-60404160301: Using Deep Learning to Generate Models for Distinguishing Underwater Targets"], 全军武器装备采购信息网 [*Whole Military Weapons and Equipment Purchase Information Net*], August 1, 2016.

"基金-61403110101-舰载系留无人机平台与缆绳系统动力学研究" ["Fund-61403110101: Ship-Based Drone Platform and Cable Dynamics Research"], 全军武器装备采购信息网 [*Whole Military Weapons and Equipment Purchase Information Net*], August 1, 2016.

"基金-61403110201-蜂群无人机数据链技术" ["Fund-61403110201: Bee Colony Drone Data Link Technology"], 全军武器装备采购信息网 [*Whole Military Weapons and Equipment Purchase Information Net*], August 1, 2016.

"基金-61403110401-化石燃料新概念无人机动力系统技术" ["Fund-61403110401: New Concepts for Drone Fossil Fuel Mechanical Systems"], 全军武器装备采购信息网 [*Whole Military Weapons and Equipment Purchase Information Net*], April 11, 2017.

"基金-61404130305-基于深度学习的雷达目标识别技术" ["Fund-61404130305: Radar Target Distinguishing Using Deep Learning"], 全军武器装备采购信息网 [*Whole Military Weapons and Equipment Purchase Information Net*], April 11, 2017.

Fung, Courtney J., "China and the Responsibility to Protect: From Opposition to Advocacy," United States Institute of Peace, June 8, 2016. As of July 2, 2018: https://www.usip.org/publications/2016/06/china-and-responsibility-protect-opposition -advocacy

Future of Life Institute, "Autonomous Weapons: An Open Letter from AI & Robotics Researchers," July 28, 2015. As of August 30, 2018: https://futureoflife.org/open-letter-autonomous-weapons/?cn-reloaded=1

Gambhir, Mahak, and Vishal Gupta, "Recent Automatic Text Summarization Techniques: A Survey," *Artificial Intelligence Review*, Vol. 47, No. 1, 2017, pp. 1–66.

葛卫丽 [Ge Weili] and 田春鸣 [Tian Chunming], "'指挥自动化系统与技术'课程建设与现状分析" ["Building a Course in and Analyzing the State of 'Command Automation System Technology'"], 武警技术学院学报 [*Journal of the People's Armed Police Technical Academy*], No. 2, 1998. As of July 8, 2018:
http://kns55.en.eastview.com/kcms/detail/detail.aspx?recid=&FileName=WJGX199802005&DbName=CJFD1998&DbCode=CJFD&uid=WEE2cU9XZDE2YVM1YnFrPQ

Geist, Edward M., "It's Already Too Late to Stop the AI Arms Race—We Must Manage It Instead," *Bulletin of the Atomic Scientists*, Vol. 72, No. 5, 2016, pp. 318–321. As of July 18, 2019:
DOI: 10.1080/00963402.2016.1216672

Geist, Edward, and Andrew J. Lohn, *How Might Artificial Intelligence Affect the Risk of Nuclear War?* Santa Monica, Calif.: RAND Corporation, PE-296-RC, 2018. As of September 5, 2018:
https://www.rand.org/pubs/perspectives/PE296.html

Gigova, Radina, "Who Vladimir Putin Thinks Will Rule the World," *CNN*, September 2, 2017. As of June 25, 2018:
https://www.cnn.com/2017/09/01/world/putin-artificial-intelligence-will-rule-world/index.html

"金戈铁马骋沙场 中国装甲展雄风" ["Golden Axes and Iron Horses Gallop onto the Battlefield, China's Armored Vehicles Become a Powerful Wind"], *NORINCO*, August, 2015.

Goodwin, Thomas G., and Don Shipley, "Robots Conquer DARPA Grand Challenge," DARPA News Release, October 8, 2005.

"Google Achieves AI 'Breakthrough' by Beating Go Champion," *BBC News*, January 27, 2016. As of June 29, 2018:
https://www.bbc.com/news/technology-35420579

Gross, Kerianne H., Matthew A. Clark, Jonathan A. Hoffman, Eric D. Swenson, and Aaron W. Fifarek, "Run-Time Assurance and Formal Methods Analysis Nonlinear System Applied to Nonlinear System Control," *Journal of Aerospace Information Systems*, Vol. 14, No. 4, 2017, pp. 232–246. As of July 18, 2019:
DOI: 10.2514/1.I010471.

Grossman, Derek, and Michael S. Chase, "Why Xi Is Purging the Chinese Military," *The National Interest*, April 15, 2016, As of July 8, 2018:
http://nationalinterest.org/feature/why-xi-purging-the-chinese-military-15795

顾云涛 [Gu Yuntao], "人工智能技术在武器投放系统中的应用" ["Application of Artificial Intelligence Technology on Weapon Delivery Systems"], 现代导航 [*Modern Navigation*],

2013, pp. 452–456. As of July 18, 2019:
http://www.ixueshu.com/document/7a8cb13c74bcebc9318947a18e7f9386.html#original

Gubrud, Mark, "Why Should We Ban Autonomous Weapons? To Survive," *IEEE Spectrum*, 2016. As of August 29, 2018:
https://spectrum.ieee.org/automaton/robotics/military-robots/why-should-we-ban-autonomous-weapons-to-survive

Gunning, David, "Explainable Artificial Intelligence (XAI)," Defense Advanced Research Projects Agency (DARPA), n.d. As of September 10, 2018:
https://www.darpa.mil/program/explainable-artificial-intelligence

"Half of Russian PhD Students Want to Move Abroad," *Moscow Times*, April 4, 2018. As of July 6, 2018:
https://themoscowtimes.com/news/half-russian-phd-students-want-move-abroad-61050

Hambling, David, "If Drone Swarms Are the Future, China May Be Winning," *Popular Mechanics*, December 23, 2016. As of July 2, 2018.
https://www.popularmechanics.com/military/research/a24494/chinese-drones-swarms/

———, "Iran's 'New' Anti-Missile Artillery," *WIRED*, May 27, 2009. As of June 27, 2018:
https://www.wired.com/2009/05/irans-new-anti-missile-artillery/

Harding, Nick, "'Deepfake' Videos Produced by Russian-Linked Trolls Are the Latest Weapon in Fake News War, Official Monitors Warn," *The Telegraph*, May 26, 2018. As of July 6, 2018:
https://www.telegraph.co.uk/news/2018/05/26/deepfake-videos-produced-russian-linked-trolls-latest-weapon/

Hawley, John K., "Looking Back at 20 Years of MANPRINT on Patriot: Observations and Lessons," Army Research Laboratory, ARL-SR-0158, September 2007. As of August 30, 2018:
http://www.arl.army.mil/arlreports/2007/ARL-SR-0158.pdf

He Kaiming, Xiangyu Zhang, Shaoqing Ren, and Jian Sun, "Delving Deep into Rectifiers: Surpassing Human-Level Performance on ImageNet Classification," *Proceedings of the 2015 International Conference on Computer Vision*, Santiago, Chile, December 2015.

Hedberg, Sara Reese, "DART: Revolutionizing Logistics Planning," *IEEE Intelligent Systems*, Vol. 17, 2002, pp. 81–83. As of July 18, 2019:
DOI: 10.1109/MIS.2002.1005635

Hennessey, Meghan, "Clearpath Robotics Takes Stance Against 'Killer Robots,'" August 13, 2014. As of August 30, 2018:
https://www.clearpathrobotics.com/2014/08/clearpath-takes-stance-against-killer-robots/

Heyns, Christof, "Autonomous Weapons Systems and Human Rights Law," presentation at the informal expert meeting organized by the state parties to the Convention on Certain Conventional Weapons, Geneva, May 2014.

Hochreiter, Sepp, and Jürgen Schmidhuber, "Long Short-Term Memory," *Neural Computation*, Vol. 9, No. 8, 1997, pp. 1735–1780.

Hoffman, David, "I Had a Funny Feeling in My Gut," *Washington Post*, February 10, 1999. As of July 6, 2018:
https://www.washingtonpost.com/wp-srv/inatl/longterm/coldwar/shatter021099b.htm

Horowitz, Michael C., "Artificial Intelligence, International Competition, and the Balance of Power," *Texas National Security Review*, Vol. 1, No. 3, May 2018, pp. 36–57.

———, "Public Opinion and the Politics of the Killer Robots Debate," *Research and Politics*, Vol. 3, No. 1, January–March 2016, pp. 1–8.

Horowitz, Michael, and Paul Scharre, "Meaningful Human Control in Weapon Systems: A Primer," CNAS Working Paper, Center for a New American Security, March 2015. As of September 10, 2018:
https://s3.amazonaws.com/files.cnas.org/documents/Ethical_Autonomy_Working_Paper_031315.pdf?mtime=20160906082316

HRW—*See* Human Rights Watch.

"Human Machine Touchpoints: The United Kingdom's Perspective on Human Control over Weapon Development and Targeting Cycles," Paper submitted by the United Kingdom to Group of Governmental Experts of the High Contracting Parties to the Convention on Prohibitions or Restrictions on the Use of Certain Conventional Weapons Which May Be Deemed to Be Excessively Injurious or to Have Indiscriminate Effects, August 2018. As of September 5, 2018:
https://www.unog.ch/__80256ee600585943.nsf/(httpPages)/7c335e71dfcb29d1c1258243003e8724?OpenDocument&ExpandSection=3#_Section3

Human Rights Watch and International Human Rights Clinic, *Losing Humanity: The Case Against Killer Robots*, 2012. As of September 5, 2018:
https://www.hrw.org/sites/default/files/reports/arms1112_ForUpload.pdf

———, *Mind the Gap: The Lack of Accountability for Killer Robots*, April 2015. As of September 5, 2018: https://www.hrw.org/sites/default/files/report_pdf/arms1216_web.pdf

———, *Making the Case: The Dangers of Killer Robots and the Need for a Preemptive Ban*, 2016.

"信息系统-315020301-大数据背景下基于深度学习的多源情报分析与预测技术"
["Information Systems-315020301: Technology That Uses Big Data and Machine Learning

on Multi-Source Information to Analyze and Forecast"], 全军武器装备采购信息网 [*Whole Military Weapons and Equipment Purchase Information Net*], August 5, 2016.

"The Inside Story of the Swords Armed Robot 'Pullout' in Iraq: Update," *Popular Mechanics*, September 30, 2009, As of July 8, 2018:
https://www.popularmechanics.com/technology/gadgets/a2804/4258963/

International Committee of the Red Cross, "Ethics and Autonomous Weapon Systems: An Ethical Basis for Human Control?" April 2018.

———, *Protocols Additional to the Geneva Convention of 12 August 1949*. As of August 18, 2019:
https://www.icrc.org/en/doc/assets/files/other/icrc_002_0321.pdf

Irving, Doug, "Big Data, Big Questions," *RAND Review*, October 16, 2017. As of July 9, 2018:
https://www.rand.org/blog/rand-review/2017/10/big-data-big-questions.html

"ISO/IEC JTC 1/SC 42: Artificial Intelligence," International Organization for Standardization, n.d. As of August 26, 2019:
https://www.iso.org/committee/6794475/x/catalogue/p/1/u/1/w/1/d/1

Joint Chiefs of Staff, "Gen. Selva's Q&A Session at the Brookings Institution," Brookings Foreign Policy Program, January 2016. As of September 5, 2018:
http://www.jcs.mil/Media/Speeches/Article/645038/gen-selvas-qa-session-at-the-brookings
-institution/

"Just How Much of a Threat Is Russia's Status-6 Nuclear Torpedo?" The National Interest, January 16, 2018. As of July 6, 2018:
http://nationalinterest.org/blog/the-buzz/just-how-much-threat-russias-status-6-nuclear
-torpedo-24094

"Kalashnikov Gunmaker Develops Combat Module Based on Artificial Intelligence," *TACC* [*TASS*], July 5, 2017. As of June 14, 2018:
http://tass.com/defense/954894

Kania, Elsa B., *Battlefield Singularity: Artificial Intelligence, Military Revolution, and China's Future Military Power*, Washington, D.C.: Center for New American Security, 2017.

Kaplan, Jerry, *Artificial Intelligence: What Everyone Needs to Know*, Oxford, UK: Oxford University Press, 2016.

Kavanagh, Jennifer, and Michael D. Rich, *Truth Decay: An Initial Exploration of the Diminishing Role of Facts and Analysis in American Public Life*, Santa Monica, Calif.: RAND Corporation, RR-2314-RC, 2018. As of September 11, 2018:
https://www.rand.org/pubs/research_reports/RR2314.html

Keck, Zachary, "China to Lead World in Drone Production," *The Diplomat*, May 2, 2014. As of July 2, 2018:
https://thediplomat.com/2014/05/china-to-lead-world-in-drone-production/

Keller, John, "DARPA Seeks to Improve Machine Autonomy to Enable Its Use in Safety-Critical Aircraft Applications," Military & Aerospace, July 18, 2017. As of June 29, 2018:
https://www.militaryaerospace.com/articles/2017/07/machine-autonomy-safety-critical-aircraft.html

Kim, Yoochul, and Minhyung Lee, "Humans Are Still Better Than AI at StarCraft—for Now," *MIT Technology Review*, November 1, 2017. As of June 25, 2018:
https://www.technologyreview.com/s/609242/humans-are-still-better-than-ai-at-starcraftfor-now/

Kozyulin, Vadim, and Albert Efimov, "Новыи Бонд—Машина С Лицензиеи На Убийство" ["The New Bond: A Machine with License to Kill"], *Индекс Безопасности* [*Security Index*], Vol. 22, No. 1, pp. 28–29. As of July 6, 2018:
http://pircenter.org/media/content/files/13/14636517760.pdf

Kozyulin, Vadim, Tom Grant, Gilles Giac, Albert Efimov, Xinping Song Xian, and Mary Wareham, "Боевые Роботы: Угрозы Учтенные Или Непредвиденные?" ["Combat Robots: Recognized or Unexpected Threats?"], *Индекс Безопасности* [*Security Index*], Vol. 22, Nos. 3–4, 2016, p. 96. As of July 6, 2018:
http://www.pircenter.org/media/content/files/13/14875332590.pdf

Kruglov, Alexander, Alexey Ramm, and Evgeny Dmitriev, "Средства ПВО Объединят Искусственным Интеллектом" ["Air Defense Weapons Will Be Combined with Artificial Intelligence"], *Известиыа* [*Izvestiya*], May 2, 2018. As of July 6, 2018:
https://iz.ru/733333/aleksandr-kruglov-aleksei-ramm-evgenii-dmitriev/sredstva-pvo-obediniat-iskusstvennym-intellektom

LaGrone, Sam, "Mabus: F-35 Will Be 'Last Manned Strike Fighter' the Navy, Marines 'Will Ever Buy or Fly," *USNI News*, April 15, 2015. As of June 28, 2018:
https://news.usni.org/2015/04/15/mabus-f-35c-will-be-last-manned-strike-fighter-the-navy-marines-will-ever-buy-or-fly

Landmine and Cluster Munition Monitor, "China Cluster Munition Ban Policy," *The Monitor* July 21, 2016. As of July 2, 2018:
http://www.the-monitor.org/en-gb/reports/2016/china/cluster-munition-ban-policy.aspx

"Laser Weapons," *Jane's Strategic Weapons Systems*, July 24, 2015.

Latiff, Robert H., *Future War: Preparing for the New Global Battlefield*, New York: Alfred A. Knopf, 2017.

Le, Quoc V., and Mike Schuster, "A Neural Network for Machine Translation, at Production Scale," *Google AI Blog*, September 27, 2016. As of June 25, 2018:
https://ai.googleblog.com/2016/09/a-neural-network-for-machine.html

Leveson, Nancy G., *Engineering a Safer World: Systems Thinking Applied to Safety*, Cambridge, Mass.: MIT Press, 2011.

Levine, Sergey, Peter Pastor, Alex Krizhevsky, and Deirdre Quillen, "Learning Hand-Eye Coordination for Robotic Grasping with Deep Learning and Large-Scale Data Collection," *International Journal of Robotics Research*, Vol. 37, Nos. 4–5, 2017, pp. 421–436.

李倩影 [Li Qianying], "人工智能专家联名呼吁联合国禁止'杀手机器人" ["AI Experts Call on the UN to Ban 'Killer Robots'"], *Xinhua News*, August 21, 2017. As of July 3, 2018:
http://www.81.cn/rd/2017-08/21/content_7726562.htm

Lieberthal, Kenneth, and Jisi Wang, *Addressing U.S.-China Strategic Distrust*, Washington, D.C.: The Brookings Institution John L. Thronton China Center. March 2012. As of July 2, 2018:
http://yahuwshua.org/en/Resource-584/0330_china_lieberthal.pdf

"轻型通用无人平台动力源集成化技术" ["Light Integrated Power Generation Technology for Drones"], 全军武器装备采购信息网 [*Whole Military Weapons and Equipment Purchase Information Net*], November 7, 2017.

Lin, Jeffrey, and P. W. Singer, "China's New Military Robots Pack More Robots Inside (StarCraft Style)," *Popular Science*, November 11, 2014. As of July 2, 2018:
https://www.popsci.com/blog-network/eastern-arsenal/chinas-new-military-robots-pack-more-robots-inside-starcraft-style#page-2

———, "New Chinese Laser Weapon Stars on TV," *Popular Science*, November 25, 2015. As of July 2, 2018:
https://www.popsci.com/new-chinese-laser-weapon-stars-on-tv#page-2

———, "Meet China's Sharp Sword, a Stealth Drone That Can Likely Carry 2 Tons of Bombs," *Popular Science*, January 18, 2017. As of July 2, 2018:
https://www.popsci.com/china-sharp-sword-lijian-stealth-drone

林岩峰 [Lin Yanfeng], "人工智能将取代战场指挥官" ["Will Artificial Intelligence Replace the Battlefield Commander?"], Chinese Ministry of National Defense, June 23, 2017, As of July 8, 2018:
http://www.mod.gov.cn/jmsd/2017-06/23/content_4783506.htm

刘航 [Liu Hang], ed., "未来战场新锐：军用机器人" ["The Future Battlefield Is New: Military Robots"], *People's Daily*, August 27, 2016, As of July 8, 2018:
http://www.81.cn/2017jj90/2016-08/27/content_7669636.htm

Lockheed Martin, "Long Range Anti-Ship Missile (LRASM)," YouTube, May 2, 2016. As of June 28, 2018:
https://www.youtube.com/watch?v=h449oIjg2kY

Lubold, Gordon, and Jeremy Page, "American Military Aircraft Targeted by Lasers in Pacific Ocean, U.S. Officials Say," *Wall Street Journal*, June 21, 2018. As of July 2, 2018:
https://www.wsj.com/articles/american-military-aircraft-targeted-by-lasers-in-pacific -ocean-u-s-officials-say-1529613999?utm_source=Sailthru&utm_medium=email&utm _campaign=ebb%2022.06.18&utm_term=Editorial%20-%20Early%20Bird%20Brief

Lucas, Louise, Nicolle Liu, and Yingzhi Yang, "China Chatbot Goes Rogue: 'Do You Love the Communist Party?' 'No,'" *Financial Times*, August 2, 2017. As of July 8, 2018:
https://www.ft.com/content/e90a6c1c-7764-11e7-a3e8-60495fe6ca71?mhq5j=e1

Lynch, Shana, "Andrew Ng: Why AI Is the New Electricity," *Insights by Stanford Business*, Stanford Graduate School of Business, March 11, 2017. As of June 25, 2018:
https://www.gsb.stanford.edu/insights/andrew-ng-why-ai-new-electricity

"重点实验室基金-61421040105-面向多无人平台的自组织网络理论与关键技术" ["Major Laboratory Fund-61421040105: Network Theory and Critical Technology for Self-Organizing Among Multiple Drone Platforms"], 全军武器装备采购信息网 [*Whole Military Weapons and Equipment Purchase Information Net*], May 19, 2017.

"重点实验室基金-61422150101-面向水中无人航行器的人工智能方法" ["Major Laboratory Fund-61422150101: Navigation Devices for USVs"], 全军武器装备采购信息网 [*Whole Military Weapons and Equipment Purchase Information Net*], May 19, 2017.

"重点实验室基金-61422150307-新概念水中无人航行器系统开发" ["Major Laboratory Fund-61422150307: Using New Concepts for USV Navigation"], 全军武器装备采购信息网 [*Whole Military Weapons and Equipment Purchase Information Net*], May 19, 2017.

"重点实验室基金-61423011001-异构多无人机协同任务分配、资源优化和路径规划系统" ["Major Laboratory Fund-614230011001: Mission Distribution, Resource Optimization, and Route Planning in Large, Varied Drone Swarms"], 全军武器装备采购信息网 [*Whole Military Weapons and Equipment Purchase Information Net*], May 19, 2017.

"重点实验室基金-61423010902-无人机数据链抗干扰方法研究" ["Major Laboratory Fund-61423010902: Research into Interference Resistant Drone Data Links"], 全军武器装备采购信息网 [*Whole Military Weapons and Equipment Purchase Information Net*], May 19, 2017.

"重点实验室基金-61425030202-微波成像自动识别技术" ["Major Laboratory Fund-61425030202: Autonomous Identification in Microwave Images"], 全军武器装备采购信息网 [*Whole Military Weapons and Equipment Purchase Information Net*], May 19, 2017.

"重点实验室基金-6142A010103-多源遥感大数据支持下的地物光谱匹配新模型研究" ["Major Laboratory Fund-6142A010103: Research on Using Multi-Source Big Data to Support Spectrum Match Modeling"], 全军武器装备采购信息网 [*Whole Military Weapons and Equipment Purchase Information Net*], May 19, 2017.

McCarthy, John, and Patrick J. Hayes, "Some Philosophical Problems from the Standpoint of Artificial Intelligence," Stanford University, 1969. As of January 31, 2019:
http://jmc.stanford.edu/articles/mcchay69/mcchay69.pdf

McCarthy, John, Marvin L. Minsky, Nathaniel Rochester, and Claude E. Shannon, "A Proposal for the Dartmouth Summer Research Project on Artificial Intelligence, August 31, 1955," *AI Magazine*, Vol. 27, No. 4, 2006, p. 12. As of May 27, 2018:
https://www.aaai.org/ojs/index.php/aimagazine/article/download/1904/1802

McDuffee, Allen, "Black Hawk Drone: Army's Iconic Helicopter Goes Pilotless," *WIRED*, April 30, 2014. As of June 28, 2018:
https://www.wired.com/2014/04/black-hawk-drone/

McGrath, James R., "Twenty-First Century Information Warfare and the Third Offset Strategy," *Joint Force Quarterly*, Vol. 82, No. 3, 2016, pp. 16–23. As of August 31, 2018:
http://ndupress.ndu.edu/Portals/68/Documents/jfq/jfq-82/jfq-82.pdf?ver=2016-07-08-153513-167

Meier, Michael, "U.S. Delegation Opening Statement (as delivered)," U.S. Mission Geneva Convention on Certain Conventional Weapons (CCW) Informal Meeting of Experts on Lethal Autonomous Weapons Systems (LAWS), April 2016. As of September 5, 2018:
https://geneva.usmission.gov/2016/04/11/laws/

Metz, Cade, "In a Huge Breakthrough, Google's AI Beats a Top Player at the Game of Go," *WIRED*, January 27, 2016b. As of June 25, 2018:
https://www.wired.com/2016/01/in-a-huge-breakthrough-googles-ai-beats-a-top-player-at-the-game-of-go/

———, "In Two Moves, AlphaGo and Lee Sedol Redefined the Future," *WIRED*, March 16, 2016a. As of June 25, 2018:
https://www.wired.com/2016/03/two-moves-alphago-lee-sedol-redefined-future/

Mewett, Christopher, "Understanding War's Enduring Nature Alongside Its Changing Character," *War on the Rocks*, January 21, 2014. As of September 7, 2018:
https://warontherocks.com/2014/01/understanding-wars-enduring-nature-alongside-its-changing-character/

Meyer, Tom J., "Active Protective Systems: Impregnable Armor or Simply Enhanced Survivability?" *ARMOR*, May–June 1998. As of July 6, 2018:
https://fas.org/man/dod-101/sys/land/docs/3aps98.pdf

军事科学院军事战略研究部 [Military Science Institute, Military Strategy Research Department], 战略学 [*The Science of Military Strategy*], ed. 孙兆利 [Sun Zhaoli], Beijing: Military Science Publishing House, 2013.

Mitchell, Billy, "'No Longer Experimental'—DIUx Becomes DIU, Permanent Pentagon Unit," *FEDSCOOP*, August 9, 2018. As of August 22, 2019:
https://www.fedscoop.com/diu-permanent-no-longer-an-experiment/

Mizintsev, Mikhail Evgenyevich, "Национальный Центр Управления Обороной Российской Федерации" ["National Center for Defense Management of the Russian Federation"], Министерство Обороны Российской Федерации [Ministry of Defense of the Russian Federation], n.d. As of July 6, 2018:
https://structure.mil.ru/structure/ministry_of_defence/details.htm?id=11206@egOrganization

Mizokami, Kyle, "China Is Experimenting with Remote Controlled Tanks," *Popular Mechanics*, March 21, 2018. As of July 2, 2018:
https://www.popularmechanics.com/military/weapons/a19544755/china-is-experimenting-with-remote-controlled-tanks/

———, "Phalanx: The US Navy's Last-Ditch Automated Air Defense System," *Popular Mechanics*, April 14, 2016. As of June 26, 2018:
https://www.popularmechanics.com/military/weapons/a20392/phalanx-the-us-navys-last-ditch-air-defense-system/

———, "The U.S. Navy Just Got the World's Largest Uncrewed Ship," *Popular Mechanics*, February 5, 2018. As of June 28, 2018:
https://www.popularmechanics.com/military/navy-ships/a16573306/navy-accept-delivery-actuv-sea-hunter/

"某型无人机用机载电池" ["Modeling Drone Batteries"], 全军武器装备采购信息网 [*Whole Military Weapons and Equipment Purchase Information Net*], July 31, 2017.

Moravec, Hans, *Mind Children*, Cambridge, Mass.: Harvard University Press, 1988.

Morgan, M. G., B. Fischhoff, A. Bostrom, and C. J. Atman, *Risk Communication: A Mental Models Approach*, New York: Cambridge University Press, 2002.

Mozur, Paul, "Internet Users in China Expect to Be Tracked. Now, They Want Privacy," *New York Times*, January 4, 2018. As of July 3, 2018:
https://www.nytimes.com/2018/01/04/business/china-alibaba-privacy.html

Mulvehill, Alice M., and Joseph A. Caroli, "JADE: A Tool for Rapid Crisis Action Planning," paper presented at the Command and Control Research and Technology Symposium, United States Naval War College, Newport, RI, 1999. As of June 27, 2018:
http://www.dtic.mil/dtic/tr/fulltext/u2/a458570.pdf

Mulvehill, Alice M., Clinton Hyde, and Dave Ranger, "Joint Assistant for Deployment and Execution (JADE)," Air Force Research Laboratory Technical Report: *AFRL-IF-RS-TR-2001-171*, August 2001. As of June 27, 2018:
http://www.dtic.mil/get-tr-doc/pdf?AD=ADA398021

Muolo, Danielle, "Why Go Is So Much Harder for AI to Beat Than Chess," *Business Insider*, March 10, 2016. As of May 27, 2018:
http://www.businessinsider.com/why-google-ai-game-go-is-harder-than-chess-2016-3

全国信息安全标准化技术委员会 [National Information Security Standardization Technical Committee], 信息安全技术 个人信息安全规范 [Personal Information Security Information Security Technology Standards], January 24, 2018. As of July 3, 2018:
https://www.tc260.org.cn/front/postDetail.html?id=20180124211617

"海军创新-30201050110-基于深度学习的海洋遥感目标信息挖掘技术" ["Naval Innovation-30201050110: Remote Sensing and Target Information Excavation with Deep Learning"], 全军武器装备采购信息网 [*Whole Military Weapons and Equipment Purchase Information Net*], March 31, 2017.

"海军创新-30201050111-基于人工智能的图像处理和舰船目标识别技术" ["Naval Innovation-30201050111: AI Image Processing and Warship Target Distinguishing Technology"], 全军武器装备采购信息网 [*Whole Military Weapons and Equipment Purchase Information Net*], March 31, 2017.

"海军创新-30201050404-基于大数据的中远海海上目标信息处理技术" ["Naval Innovation-30201050404: Technology to Use Big Data to Process Information on Targets in Middle and Far Seas"], 全军武器装备采购信息网 [*Whole Military Weapons and Equipment Purchase Information Net*], March 31, 2017.

"海军创新-30202021401-基于大数据的水声探测与识别技术" ["Naval Innovation-30202021401: Using Big Data to Find and Distinguish Sonar Targets"], 全军武器装备采购信息网 [*Whole Military Weapons and Equipment Purchase Information Net*], March 31, 2017.

"海军创新-30205020708-基于大数据分析的建造过程智能管控技术" ["Naval Innovation-30205020708: Big Data Technology to Analyze and Control the Process of Construction"], 全军武器装备采购信息网 [*Whole Military Weapons and Equipment Purchase Information Net*], March 31, 2017.

"海军预研-基于大数据的卫星信息数据挖掘技术" ["Naval Research: Big Data Satellite Image Data Mining Technology"], 全军武器装备采购信息网 [*Whole Military Weapons and Equipment Purchase Information Net*], August 1, 2016.

"海军预研-复杂电磁环境下无人机抗干扰诱骗技术" ["Naval Research: Jamming and Spoofing-Resistant Technologies for Drones in Complex Electromagnetic Environments"], 全军武器装备采购信息网 [*Whole Military Weapons and Equipment Purchase Information Net*], August 1, 2016.

"海军预研-船舶大数据挖掘与处理技术" ["Naval Research: Mining and Processing Shipping Big Data"], 全军武器装备采购信息网 [*Whole Military Weapons and Equipment Purchase Information Net*], August 1, 2016.

"海军预研-船用无人飞行器" ["Naval Research: Shipboard UAVs"], 全军武器装备采购信息网 [*Whole Military Weapons and Equipment Purchase Information Net*], August 1, 2016.

Nekrasov, Mikhail, "Russian Military Robots in Action," Russia Beyond, January 11, 2017. As of July 6, 2018:
https://www.rbth.com/multimedia/video/2017/01/11/russian-military-robots-in-action_678343

Ng, Alfred, "Microsoft Is Building a Smart Antivirus Using 400 Million PCs," *CNET*, June 27, 2018. As of June 28, 2018:
https://www.cnet.com/news/microsoft-build-smart-antivirus-using-400-million-computers-artificial-intelligence/

Nuclear Threat Initiative, "Report Warns of Potential State Bioweapons Programs," August 10, 2010. As of July 2, 2018:
http://www.nti.org/gsn/article/report-warns-of-potential-state-bioweapons-programs/

———, "China Biological Chronology," August 2013. As of July 2, 2018:
http://www.nti.org/media/pdfs/china_biological_2.pdf.

———, "U.S. Alleges Continued WMD Development in China," September 15, 2016. As of July 2, 2017:
http://www.nti.org/gsn/article/us-alleges-continued-wmd-development-in-china/

"Океанская Многоцелевая Система 'Статус-6' (Kanyon)" ["Ocean Multi-Purpose System 'Status-6' (Kanyon)"], *Новости ВПК* [*Novosti VPK*], November 27, 2016. As of July 6, 2018:
https://vpk.name/library/f/status-6.html

Office of the Secretary of Defense, *Unmanned Systems Roadmap (2007–2032)*, Washington, D.C.: Department of Defense, December 2007. As of August 30, 2018:
https://my.nps.edu/documents/106607930/106914584/UxS+SD+Roadmap+2007.pdf/626883d5-b93d-4fac-874d-55238d32390a

Oliker, Olga, "Russia's New Military Doctrine: Same as the Old Doctrine, Mostly," *Washington Post*, January 15, 2015. As of July 6, 2018: https://www.washingtonpost.com/news/monkey-

cage/wp/2015/01/15/russias-new-military
-doctrine-same-as-the-old-doctrine-mostly/?utm_term=.9f148b032029

OpenAI, "Attacking Machine Learning with Adversarial Examples," February 24, 2017. As of
September 5, 2018:
https://blog.openai.com/adversarial-example-research/

Operational Test and Evaluation, "Aegis Ballistic Missile Defense System," in *FY17 Ballistic
Missile Defense Systems*, 2017, pp. 291–296. As of June 26, 2018:
http://www.dote.osd.mil/pub/reports/FY2017/pdf/bmds/2017aegisbmd.pdf

Osborn, Kris, "Navy Conducts First Aerial Refueling of X47-B Carrier Launched Drone,"
Military.com, April 22, 2015. As of June 27, 2018:
https://www.military.com/defensetech/2015/04/22/navy-conducts-first-aerial-refueling-of
-x-47b-carrier-launched-drone

Osinga, Frans P. B., *Science, Strategy, and War: The Strategic Theory of John Boyd*, Delft:
Eburon Academic Publishers, 2005.

Paganini, Pierluigi, "Cyber Warfare: From Attribution to Deterrence," Infosec Institute,
October 3, 2016. As of July 6, 2016:
https://resources.infosecinstitute.com/cyber-warfare-from-attribution-to-deterrence/#gref

"Party Affiliation," Gallup, 2018. As of July 9, 2018:
https://news.gallup.com/poll/15370/party-affiliation.aspx

Paul, Kari, "The Shocking Details You Reveal About Yourself When You 'Like' Things on
Facebook," *MarketWatch*, March 25, 2018. As of August 5, 2018:
https://www.marketwatch.com/story/the-shocking-things-you-reveal-about-yourself-when
-you-like-things-on-facebook-2017-05-16

Pawlyk, Oriana, "China Leaving US Behind on Artificial Intelligence: Air Force General,"
Military.Com, July 30, 2018. As of March 5, 2019:
https://www.military.com/defensetech/2018/07/30/china-leaving-us-behind-artificial
-intelligence-air-force-general.html

Peck, Michael, "Killer Robots Using AI Could Transform Warfare. And China Might Hate
That," The National Interest, July 1, 2018, As of July 8, 2018:
http://nationalinterest.org/blog/buzz/killer-robots-using-ai-could-transform-warfare-and
-china-might-hate-24632

People's Republic of China Delegation, "The Position Paper Submitted by the Chinese
Delegation to CCW 5th Review Conference," Geneva, Switzerland, December 2016. As of
July 2, 2018:
https://www.unog.ch/80256EDD006B8954/(httpAssets)/DD1551E60648CEBBC125808A00
5954FA/$file/China's+Position+Paper.pdf

Phillips, Tom, "'Your Only Right Is to Obey': Lawyer Describes Torture in China's Secret Jails," *The Guardian*, January 23, 2017. As of July 2, 2018:
https://www.theguardian.com/world/2017/jan/23/lawyer-torture-china-secret-jails-xie-yang.

Pichai, Sundar, "AI at Google: Our Principles," *AI*, June 7, 2018. As of August 30, 2018:
https://blog.google/technology/ai/ai-principles/

Polyakova, Alina, and Spencer P. Boyer, "The Future of Political Warfare: Russia, the West, and the Coming Age of Global Competition," Foreign Policy at Brookings, March 2018. As of July 6, 2018:
https://www.brookings.edu/wp-content/uploads/2018/03/the-future-of-political-warfare.pdf

Pomerleau, Mark, "DoD Plans to Invest $600M in Unmanned Undersea Vehicles," Defense Systems, February 4, 2016. As of June 28, 2018:
https://defensesystems.com/articles/2016/02/04/dod-navy-uuv-investments.aspx

Popov, Igor, "Военные Конфликты: Взгляд За Горизонт" ["Military Conflicts: A Look Beyond the Horizon"], *Независимый Военный Обзор* [*Independent Military Review*], n.d. As of July 06, 2018:
http://www.avnrf.ru/index.php/vse-novosti-sajta/527-voennye-konflikty-vzglyad-za-gorizont

Popov, Sergey, and Oleg Falichev, "Робот Стреляет Первым" ["The Robot Shoots First"], *Новости ВПК* [*Novosti VPK*], February 23, 2016. As of July 6, 2018:
https://www.vpk-news.ru/articles/29352

President's Foreign Intelligence Advisory Board, "The Soviet 'War Scare,'" February 15, 1990, p. 45. As of July 6, 2018:
https://nsarchive2.gwu.edu//nukevault/ebb533-The-Able-Archer-War-Scare-Declassified-PFIAB-Report-Released/2012-0238-MR.pdf

Price, Emily, "AI Is Better at Diagnosing Skin Cancer Than Your Doctor, Study Finds," *Fortune*, May 30, 2018. As of June 25, 2018:
http://fortune.com/2018/05/30/ai-skin-cancer-diagnosis/

Protocols Additional to the Geneva Conventions of August 12, 1949, relating to the Protection of Victims of International Armed Conflicts (Protocol I), Article 36, June 8, 1977. As of August 23, 2019:
https://www.icrc.org/en/doc/assets/files/other/icrc_002_0321.pdf

Protocol on Blinding Laser Weapons, Geneva, Switzerland, October 13, 1995. As of July 3, 2018:
https://ihl-databases.icrc.org/applic/ihl/ihl.nsf/ART/570-4?OpenDocument

"关于预先发布"跨越险阻 2018"地面无人系统挑战赛系列活动信息的公告" ["Public Notice on 'Dangerous Crossing 2018' UGV Challenge"], 全军武器装备采购信息网 [*Whole Military Weapons and Equipment Purchase Information Net*], February 13, 2018.

"共用-41402050301-基于大数据的系统级产品故障征兆发现与故障预测研究" ["Public Use-41402050301: Research on Using Big Data to Identify Indicators of Faults and Predict Malfunctions in System-Level Equipment"], 全军武器装备采购信息网 [*Whole Military Weapons and Equipment Purchase Information Net*], April 11, 2017.

"共用-41411020301-小型长航时无人机技术" ["Public Use-41411020301: Small-Scale Long-Endurance Drone Technology"], 全军武器装备采购信息网 [*Whole Military Weapons and Equipment Purchase Information Net*], April 11, 2017.

"共用-41411030501-小型固定翼无人机密集编队飞行与防撞控制技术" ["Public Use-41411030501: Technology for Small Fixed-Wing Drones to Congregate, Fly in Formation, and Avoid Collisions"], 全军武器装备采购信息网 [*Whole Military Weapons and Equipment Purchase Information Net*], April 11, 2017.

"Putin Takes Greater Control of Russia's Defense Sector," Stratfor, September 8, 2014. As of July 06, 2018:
https://worldview.stratfor.com/article/putin-takes-greater-control-russias-defense-sector#/entry/jsconnect?client_id=644347316&target=%2Fdiscussion%2Fembed%3Fp%3D%252Fdiscussion%252Fembed%252F%26title%3DPutin%2BTakes%2BGreater%2BControl%2BOf%2BRussia%2527s%2BDefense%2BSector%26vanilla_category_id%3D1%26vanilla_identifier%3D268143%26vanilla_url%3Dhttps%253A%252F%252Fworldview.stratfor.com%252Farticle%252Fputin-takes-greater-control-russias-defense-sector

Putin, Vladimir, "Presidential Address to the Federal Assembly," March 1, 2018. As of July 6, 2018:
http://en.kremlin.ru/events/president/news/56957

Ramm, Alexey, "Новые Системы Смогут Эффективно Решать Задачи Без Участия Человека" ["New Systems Can Effectively Solve Problems Without Human Intervention"], *Известиыа* [*Izvestiya*], April 27, 2018. As of July 6, 2018:
https://iz.ru/733218/aleksei-ramm/novye-sistemy-smogut-effektivno-reshat-zadachi-bez-uchastiia-cheloveka

Ramm, Alexey, Dmitry Litovkin, and Evgeny Andreev, "На Пути К Искусственному Интеллекту" ["The Forces of Electronic Warfare Will Bring Artificial Intelligence"], *Известиыа* [*Izvestiya*] April 4, 2017. As of July 6, 2018:
https://iz.ru/news/675891

Rappert, Brian, Richard Moyes, Anna Crowe, and Thomas Nash, "The Roles of Civil Society in the Development of Standards Around New Weapons and Other Technologies of Warfare," *International Review of the Red Cross*, Vol. 94, No. 886, Summer 2012, pp. 765–785. As of July 18, 2019:
https://www.icrc.org/eng/assets/files/review/2012/irrc-886-rappert-moyes-crowe-nash.pdf

Reilly, M. B., "Beyond Video Games: New Artificial Intelligence Beats Tactical Experts in Combat Simulation," *UC Magazine*, June 27, 2016. As of March 5, 2019: https://magazine.uc.edu/editors_picks/recent_features/alpha.html

"Religious Leaders Call for a Ban on Killer Robots," PAX, December 12, 2014. As of August 30, 2018: https://www.paxforpeace.nl/stay-informed/news/religious-leaders-call-for-a-ban-on-killer -robots

Ricks, Thomas E., "Investigation Finds US Missiles Downed Navy Jet," *Washington Post*, December 11, 2004. As of June 29, 2018: http://www.washingtonpost.com/wp-dyn/articles/A56199-2004Dec10.html

Riley, Tonya, "Artificial Intelligence Goes Deep to Beat Humans at Poker," *Science*, March 3, 2017. As of June 25, 2018: http://www.sciencemag.org/news/2017/03/artificial-intelligence-goes-deep-beat-humans -poker

Robitzski, Dan, "AI Can Now Manipulate People's Movements in Fake Videos," Futurism, June 6, 2018. As of August 5, 2018: https://futurism.com/ai-can-now-manipulate-peoples-movements-in-fake-videos/

Roblin, Sebastian, "Is China's New 'Laser' Rifle the Ultimate Weapon or a Paper Tiger?" The National Interest, July 8, 2018. As of July 11, 2018: http://nationalinterest.org/blog/buzz/chinas-new-laser-rifle-ultimate-weapon-or-paper-tiger -25162

Roff, Heather, "Meaningful Human Control or Appropriate Human Judgment? The Necessary Limits of Autonomous Weapons," Briefing Paper for Delegates at the Review Conference of the Convention of Certain Conventional Weapons (CCW), Geneva: United Nations Office at Geneva (UNOG), December 2016. As of September 11, 2018: http://www.article36.org/wp-content/uploads/2016/12/Control-or-Judgment_-Understanding -the-Scope.pdf

Roff, Heather, and Richard Moyes, "Meaningful Human Control, Artificial Intelligence, and Autonomous Weapons," Briefing Paper for Delegates at the Convention on Certain Conventional Weapons (CCW) Meeting of Experts on Lethal Autonomous Weapons Systems (LAWS), Geneva, April 2016. As of September 5, 2018: http://www.article36.org/wp-content/uploads/2016/04/MHC-AI-and-AWS-FINAL.pdf

Rogoway, Tyler, "China Is Surging Forward with Its Development of Advanced Stealth Combat Drones," The Drive, February 23, 2018. As of July 2, 2018: http://www.thedrive.com/the -war-zone/18727/china-is-surging-forward-with-its-development-of-advanced-stealth -combat-drones

————, "We Finally See the Wings on Boeing's MQ-25 Drone as Details About Its Genesis Emerge," The Drive, March 13, 2018. As of June 27, 2018: http://www.thedrive.com/the-war-zone/19213/we-finally-see-the-wings-on-boeings-mq-25-drone-as-details-about-its-genesis-emerge

Roland, Alex, and Philip Shiman, *Strategic Computing: DARPA and the Quest for Machine Intelligence, 1983–1993*, Cambridge Mass.: MIT Press, 2002.

"Россия Полагается на Боевых Роботов" ["Russia Relies on Combat Robots"], *Калашников Медиа* [*Kalashnikov Media*], November 20, 2017. As of July 6, 2018: https://kalashnikov.media/news/4516591

"Russia's Approaches to the Elaboration of a Working Definition and Basic Functions of Lethal Autonomous Weapons Systems in the Context of the Purposes and Objectives of the Convention," Group of Governmental Experts of the High Contracting Parties to the Convention on Prohibitions or Restrictions on the Use of Certain Conventional Weapons Which May Be Deemed to Be Excessively Injurious or to Have Indiscriminate Effects, White Paper No. 6, April 4, 2018. As of July 6, 2018: https://www.unog.ch/80256EDD006B8954/(httpAssets)/FC3CD73A32598111C1258266002F6172/$file/CCW_GGE.1_2018_WP.6_E.pdf

"Russia's Brain Drain on the Rise over Economic Woes—Report," *Moscow Times*, January 24, 2018. As of July 6, 2018: https://themoscowtimes.com/news/russias-brain-drain-on-the-rise-over-economic-woes-report-60263

Sacks, Samm, "New China Data Privacy Standard Looks More Far Reaching Than the GDPR," Center for Strategic and International Studies, January 29, 2018. As of July 3, 2018: https://www.csis.org/analysis/new-china-data-privacy-standard-looks-more-far-reaching-gdpr

Sanger, David, *The Perfect Weapon: War, Sabotage, and Fear in the Cyber Age*, London: Scribe Publications, 2018.

Sanger, David E., "U.S. Indicts 7 Iranians in Cyberattacks on Banks and a Dam," *New York Times*, March 24, 2016. As of January 29, 2019: https://www.nytimes.com/2016/03/25/world/middleeast/us-indicts-iranians-in-cyberattacks-on-banks-and-a-dam.html

Scharre, Paul, "Autonomous Weapons and Operational Risk," Center for a New American Security, February 2016. As of September 5, 2018: https://s3.amazonaws.com/files.cnas.org/documents/CNAS_Autonomous-weapons-operational-risk.pdf?mtime=20160906080515

————, *Army of None: Autonomous Weapons and the Future of War*, New York: Norton, 2018.

Scharre, Paul, and Shawn Brimley, "20YY: The Future of Warfare," *War on the Rocks*, January 29, 2014.

Schmitt, Michael, "Autonomous Weapon Systems and International Humanitarian Law: A Reply to Critics," *Harvard Law School National Security Journal*, February 5, 2013. As of September 5, 2018:
http://harvardnsj.org/2013/02/autonomous-weapon-systems-and-international-humanitarian-law-a-reply-to-the-critics/

Schmitt, Michael N., ed., *Tallinn Manual 2.0 on the International Law Applicable to Cyber Operations*, Cambridge, UK: Cambridge University Press, 2017.

"Scimago Journal and Country Rank," *Scimago*, n.d. As of July 6, 2018:
https://www.scimagojr.com/countryrank.php?category=1702&area=1700&order=it&ord=des

Shaban, Hamza, "Amazon Employees Demand Company Cut Ties with ICE," *Washington Post*, June 22, 2018. As of August 30, 2018:
https://s2.washingtonpost.com/79112e/5b322bc2fe1ff66463949cf6/YmJvdWRyZWFcmFuZC5vcmc percent3D/24/86/1a232d78ae925a1a53d8f3c3fe3481d6

"Шамиль Алиев: торпеду 'Шквал' надо опустить на глубину" ["Shamil Aliyev: The Torpedo 'Shkval' Should Be Lowered to the Deep"], *РИА Новости* [*RIA Novosti*], September 11, 2017. As of July 6, 2018:
https://ria.ru/interview/20171109/1508467879.html

Shane, Scott, and Daisuke Wakabayashi, "The Business of War: Google Employees Protest Work for the Pentagon," *New York Times*, April 4, 2018. As of January 30, 2019:
https://www.nytimes.com/2018/04/04/technology/google-letter-ceo-pentagon-project.html

Shlapak, David A., *The Russian Challenge*, Santa Monica, Calif.: RAND Corporation, PE-250-A, 2018. As of June 5, 2018:
https://www.rand.org/pubs/perspectives/PE250.html

Shoigu, Sergei, "Шойгу Призвал Военных И Гражданских Ученых Совместно Разрабатывать Роботов И Беспилотники" ["Shoigu Urged Military and Civilian Scientists to Jointly Develop Robots and Drones"], *ТАСС* [*TASS*], March 14, 2018. As of July 6, 2018:
http://tass.ru/armiya-i-opk/5028777

Shvets, Yuri B., *Washington Station: My Life as a KGB Spy in America*, New York: Simon & Schuster, 1994.

Simon, Herbert A., *The Sciences of the Artificial Intelligence*, 3rd ed., Cambridge, Mass.: MIT Press, 1996.

Simonite, Tom, "For Superpowers, Artificial Intelligence Fuels New Global Arms Race," *WIRED*, September 8, 2017. As of July 6, 2018:

https://www.wired.com/story/for-superpowers-artificial-intelligence-fuels-new-global-arms-race/

Singh, Abhijt, "Is China Really Building Missiles with Artificial Intelligence?" *The Diplomat*, September 21, 2016. As of July 2, 2018:
https://thediplomat.com/2016/09/is-china-really-building-missiles-with-artificial-intelligence/

Srivastava, Anupam, "China's Export Controls: Can Beijing's Actions Match Its Words?" Arms Control Association, November 1, 2015. As of July 2, 2018:
https://www.armscontrol.org/act/2005_11/NOV-China

"Startup Investment & Innovation in Emerging Europe," *East-West Digital News*, Version 1, February 2018. As of July 06, 2018:
http://cee.ewdn.com

国务院 中华人民共和国 [State Council, People's Republic of China], "国务院关于印发新一代人工智能发展规划的通知" ["State Council Notice on the Issuance of the Next Generation Artificial Intelligence Development Plan"], Beijing, China, July 20, 2017. As of July 3, 2018:
http://www.gov.cn/zhengce/content/2017-07/20/content_5211996.htm

StratoEnergetics, Buenos Aires Event, "Slaughterbots," YouTube, November 12, 2017. As of September 5, 2018:
https://www.youtube.com/watch?v=9CO6M2HsoIA

Subbaraman, Nidhi, "After Two Historic Carrier Landings, Navy's X47-B Drone Scrubs a Third," *NBC News*, July 11, 2013. As of June 27, 2018:
https://www.nbcnews.com/technolog/after-two-historic-carrier-landings-navys-x-47b-drone-scrubs-6C10604261

Sullivan, Kevin, "Japanese Ship Downs U.S. Plane," *Washington Post*, June 5, 1996. As of June 27, 2018:
https://www.washingtonpost.com/archive/politics/1996/06/05/japanese-ship-downs-us-plane/ba2bbbc4-5b2f-4b07-b1d2-cfc6a003e1b0/

"Suppressed in Translation," *The Economist*, March 17, 2016. As of July 3, 2018:
https://www.economist.com/news/china/21695095-how-chinese-versions-un-covenants-gloss-over-human-rights-suppressed-translation

Tegmark, Max, "Benefits and Risks of Artificial Intelligence," Future of Life, 2016. As of August 29, 2018:
https://futureoflife.org/background/benefits-risks-of-artificial-intelligence/

"国产单兵激光武器已有多款" ["There Are Already Many Chinese-Made Soldier-Usable Laser Weapons"], *Huanqiu*, December 8, 2015. As of July 2, 2018:
http://mil.huanqiu.com/photo_china/2015-12/2813143.html#p=1

Thompson, James, "Sunsetting the MQ-1 Predator: A History of Innovation," U.S. Air Force, February 20, 2018. As of July 6, 2018:
http://www.af.mil/News/Article-Display/Article/1445531/sunsetting-the-mq-1-predator-a -history-of-innovation/

天钥通航技术 [Tianyue Tonghang Technologies], "天钥'无顾虑'RF-5M 测" ["Tianyue Has 'No Misgivings' About Its RF-5M"], 全球无人机网 [*World Drone Net*], September 25, 2017. As of July 2, 2018:
http://www.81uav.cn/uav-news/201709/25/26046.html

Ticehurst, Rupert, "The Martens Clause and the Laws of Armed Conflict," *International Review of the Red Cross*, No. 317, April 30, 1997. As of August 30, 2018:
https://www.icrc.org/eng/resources/documents/article/other/57jnhy.htm

Tigner, Brooks, "Russia's 2017 Defense Spending Cut Is Not What It Seems," Atlantic Council, May 9, 2018. As of July 6, 2018:
http://www.atlanticcouncil.org/blogs/new-atlanticist/russia-s-2017-defense-spending-cut-is -not-what-it-seems

Tucker, Patrick, "The Pentagon Is Building an AI Product Factory," Defense One, April 19, 2018. As of June 28, 2018:
https://www.defenseone.com/technology/2018/04/pentagon-building-ai-product-factory/ 147594/

———, "Here's How Google Pitched AI Tools to Special Operators Last Month," Defense One, June 10, 2018. As of August 30, 2018:
https://www.defenseone.com/technology/2018/06/heres-how-google-pitched-ai-tools-special -operators-last-month/148873/

Turing, A. M., "Computing Machinery and Intelligence," *Mind*, Vol. 49, 1950, pp. 433–460. As of May 27, 2018:
https://www.csee.umbc.edu/courses/471/papers/turing.pdf

"UN Treaty Bodies and China," Human Rights in China, n.d. As of June 8, 2018:
https://www.hrichina.org/en/un-treaty-bodies-and-china

United Nations (UN), *The Universal Declaration of Human Rights*, December 10, 1948. As of September 5, 2018:
http://www.un.org/en/universal-declaration-human-rights/

United Nations Institute for Disarmament Research, "Safety, Unintentional Risk and Accidents in the Weaponization of Increasingly Autonomous Technologies," *UNIDIR Resources*, No. 5, 2016. As of September 5, 2018: http://www.unidir.org/files/publications/pdfs/safety -unintentional-risk-and-accidents-en-668.pdf

United Nations Office at Geneva, "Chair's Summary of the Discussion," Group of Governmental Experts of the High Contracting Parties to the Convention on Prohibitions or Restrictions on the Use of Certain Conventional Weapons Which May Be Deemed to Be Excessively Injurious or to Have Indiscriminate Effects, Geneva, April 2018. As of August 22, 2019:
https://www.unog.ch/80256EDD006B8954/(httpAssets)/DF486EE2B556C8A6C125827A00
488B9E/$file/Summary+of+the+discussions+during+GGE+on+LAWS+April+2018.pdf

———, *Convention on Certain Conventional Weapons*, United Nations Publication, 2014.

United Nations Office for Disarmament Affairs, "Pathways to Banning Fully Autonomous Weapons," October 23, 2017. As of January 31, 2019:
https://www.un.org/disarmament/update/pathways-to-banning-fully-autonomous-weapons/

United States, "Human-Machine Interaction in the Development, Deployment and Use of Emerging Technologies in the Area of Lethal Autonomous Weapon Systems," working paper presented at the Conference on Certain Conventional Weapons (CCW), Geneva: United Nations Office at Geneva (UNOG), August 2018. As of September 5, 2018:
https://www.unog.ch/80256EDD006B8954/(httpAssets)/D1A2BA4B7B71D29FC12582F600
4386EF/$file/2018_GGE+LAWS_August_Working+Paper_US.pdf

U.S. Census Bureau, "FINC-01. Selected Characteristics of Families by Total Money Income," 2018. As of July 12, 2019:
https://www.census.gov/data/tables/time-series/demo/income-poverty/cps-finc/finc-01.html

———, "QuickFacts: United States," 2018. As of July 9, 2018:
https://www.census.gov/quickfacts/fact/table/US/PST045217#qf-headnote-b

U.S. Chamber of Commerce, *Made in China 2025: Global Ambitions Built on Local Protections*, 2017. As of July 2, 2018:
https://www.uschamber.com/sites/default/files/final_made_in_china_2025_report_full.pdf

U.S. Congress, Senate Committee on Armed Services, "Hearing to Consider the Nomination of General Paul J. Selva, USAF, for Reappointment to the Grade of General and Reappointment to Be Vice Chairman of the Joint Chiefs of Staff," Washington, D.C., July 18, 2017. As of September 5, 2018:
https://www.armed-services.senate.gov/hearings/17-07-18-nomination_--selva

U.S. Department of Defense, *Law of War Manual*, June 2015. As of September 5, 2018:
https://dod.defense.gov/Portals/1/Documents/pubs/Law-of-War-Manual-june-2015.pdf

———, "Remarks by Deputy Secretary Work on Third Offset Strategy," Brussels, April 28, 2016. As of July 18, 2019:
https://www.defense.gov/News/Speeches/Speech-View/Article/753482/remarks-by-deputy
-secretary-work-on-third-offset-strategy/

U.S. Department of the Treasury, *2016 Report to Congress on China's WTO Compliance*, January, 2017. As of July 2, 2018:
https://ustr.gov/sites/default/files/2016-China-Report-to-Congress.pdf

Vanian, Jonathan, "Google Plans Big AI Push in Asia," *Fortune*, December 13, 2017. As of August 30, 2018:
http://fortune.com/2017/12/13/google-china-artificial-intelligence/

Voelcker, John, "Autonomous Vehicles Complete DARPA Urban Challenge," *SPECTRUM*, November 1, 2007. As of August 29, 2018:
https://spectrum.ieee.org/transportation/advanced-cars/autonomous-vehicles-complete-darpa-urban-challenge

Wagener, Gerard, and Alexandre Dulaunoy, "Torinji: Automated Exploitation Malware Targeting Tor Users," Radu State University of Luxembourg, May 24, 2009. As of August 5, 2018:
https://arxiv.org/pdf/1208.2877.pdf

Wakabayashi, Daisuke, and Scott Shane, "Google Will Not Renew Pentagon Contract That Upset Employees," *New York Times*, June 1, 2018. As of June 28, 2018:
https://www.nytimes.com/2018/06/01/technology/google-pentagon-project-maven.html

———, "Google to Quit Pentagon Work That Riled Staff," *New York Times*, June 2, 2018. As of August 30, 2018:
https://www.nytimes.com/2018/06/01/technology/google-pentagon-project-maven.html

Wassmuth, Daniel, and David Blair, "Loyal Wingman, Flocking, and Swarming: New Models of Distributed Airpower," *War on the Rocks*, February 21, 2018. As of June 28, 2018:
https://warontherocks.com/2018/02/loyal-wingman-flocking-swarming-new-models-distributed-airpower/

Welsh, Sean, "China's Shock Call for Ban on Lethal Autonomous Weapon Systems," *IHS Jane's Defense Weekly*, April 16, 2018.

"Где Найти Интеллект для Бизнеса," ["Where to Find Intelligence for Business"], SAP Planet, 2017. As of August 23, 2019:
http://sapplanet.ru/journals/sap-planet/2017/1/gde-najti-intellekt-dlya-biznesa.html

"'Whoever Leads in AI Will Rule the World': Putin to Russian Children on Knowledge Day," *RT International*, September 1, 2017. As of June 24, 2018:
https://www.rt.com/news/401731-ai-rule-world-putin/

Wong, Kelvin, "China's CASC Unveils D3000 Unmanned Oceanic Combat Vessel Concept," *Jane's International Defense Review*, September 19, 2017.

———, "China's Yunzhou Tech Performs Swarming USV Demonstration," *Jane's International Defense Review*, June 5, 2018.

———, "Image Emerges of China's Stealthy Dark Sword UCAV," *Jane's Defense Weekly*, June 7, 2018.

"World Summit of Nobel Peace Laureates: Final Declaration," Pressenza, December 14, 2014. As of August 30, 2018:
https://www.pressenza.com/2014/12/world-summit-nobel-peace-laureates-final-declaration/

World Trade Organization, "Disputes by Member," 2018. As of June 7, 2018:
https://www.wto.org/english/tratop_e/dispu_e/dispu_by_country_e.htm#top

夏文军 [Xia Wenjun], ed., 军队指挥学教程 [*Lectures on the Science of Military Command*], Beijing: Military Science Publishing House, 2012.

许述 [Xu Shu], "机器人'叛乱' 让它当战士还有多远" ["Robot 'Rebellion' Makes Robot Soldiers Even More Unlikely"], *People's Daily*, August 13, 2017, As of July 8, 2018:
http://www.81.cn/jkhc/2017-08/13/content_7713111.htm

Yashin, S. V., V. N. Denisov, O. V. Sayapin, and L. V. Makarciev, "Апробация Модели Применения Межвидовой Группировки Войск (Сил) На Стратегическом Командно-Штабном Учении 'Кавказ-2016'" ["Praise for the Model Taking the Inter-Service Correlation of Forces (Strength) in the Strategic Command Post Exercise 'Kavkaz-2016'"], *Военнаыа Мисл* [*Voennaya Misl*], No. 2, February 2018, pp. 28–32. As of July 6, 2018:
https://dlib.eastview.com/browse/doc/50730567?searchLink=%2Fsearch%2Fsimple

袁艺 [Yuan Yi], 人工智能将指挥未来战争? ["Will AI Command the Wars of Tomorrow?"], 中国国防日报 [*China Defense News*], January 12, 2017. As of July 2, 2018:
http://www.mod.gov.cn/jmsd/2017-01/12/content_4769771.htm

Yudkowsky, Eliezer, "Artificial Intelligence as a Positive and Negative Factor in Global Risk," in *Global Catastrophic Risks*, ed. Nick Bostrom and Milan M. Ćirković, New York: Oxford University Press, 2011, pp. 308–345.

Yue Pan, "China's $798B Government Funds Redraw Investment Landscape, Here Are the Largest Funds You Must Know," China Money Network, October 31, 2017. As of September 12, 2018:
https://www.chinamoneynetwork.com/2017/10/31/chinas-798b-government-funds-redraw-investment-landscape-largest-funds-must-know

Zhang Bo, "Computer Vision vs. Human Vision," *Proceedings of the 9th IEEE International Conference on Cognitive Informatics*, ICCI 2010, July 7–9, 2010, Beijing, China, pp. 1–3.

张乾 [Zhang Gan], "周志华：为什么要抵制韩国大学研发自主武器" ["Zhou Zhihua: Why We Should Boycott Korean Universities Developing Autonomous Weapons"], *AI Era*, April 8, 2018. Translation by Jeffrey Ding. As of July 3, 2018:
https://mp.weixin.qq.com/s/nNOAXkYlLjGxJ-QxrQO4PQ

张梦然 [Zhang Mengran], "联合国明年讨论致命性自主武器系统 管控'杀手机器人" ["Next Year the UN Will Discuss Lethal Autonomous Weapons Systems and Control 'Killer Robots'"], 科技日报 [*Science and Technology Daily*], December 24, 2016. As of July 3, 2018:
http://www.stdaily.com/index/kejixinwen/2016-12/24/content_494885.shtml

赵亮 [Zhao Liang], "新武器法律审查问题初探" ["An Early Legal Exploration of New Weapons"], *Journal of the Xi'an Politics Institute of the PLA*, Vol. 23, No. 2, 2010. As of July 3, 2018:
http://kns55.en.eastview.com/kcms/detail/detail.aspx?recid=&FileName=XAZX201002030&DbName=CJFD2010&DbCode=CJFD&uid=WEE2cU9OTkY1UG5sb1RZPQ

赵汀阳 [Zhao Tingyang], "人工智能'革命'的'近忧'和'远虑'—种伦理学和存在论的分析" ["'Short-Term Problems' and 'Long-Term Worries' of the AI 'Revolution'—An Ethical and Existential Analysis"], 哲学动态杂志 [*Philosophical Trends*], Vol. 1, No. 4, June 6, 2018, pp. 5–12. As of July 3, 2018:
https://mp.weixin.qq.com/s?__biz=MzAwNDYzNDI2Ng==&mid=2651908570&idx=1&sn=1af439bb4f763581dbafbc9f999e12d1&chksm=80ccaa2fb7bb23393ac527c1f25576c8b050df112147028824eb408b537b167d9b9d40519f51&mpshare=1&scene=1&srcid=06063yjlFG3wokUvacc6Hlv4&pass_ticket=%2BATm0dDoPkiEZhWNx6F%2BBQDYLCaq0e%2BC5WiRETs1EVBRH4FCdiokIpQuXZkUSV#rd

朱启超 [Zhu Qichao] and 王婧凌 [Wang Jingling], "人工智能叩开智能化战争大门" ["AI Has Knocked Open the Door to Smart War"], 解放军报 [*People's Liberation Army Daily*], January 23, 2017. As of July 2, 2018:
http://www.mod.gov.cn/jmsd/2017-01/23/content_4770692.htm